Mechanical Engineering

Mechanical Engineering

Adrian Fletcher

www.willfordpress.com

Published by Willford Press,
118-35 Queens Blvd., Suite 400,
Forest Hills, NY 11375, USA

ISBN: 978-1-64728-335-3

Cataloging-in-Publication Data

Mechanical engineering / Adrian Fletcher.
p. cm.
Includes bibliographical references and index.
ISBN 978-1-64728-335-3
1. Mechanical engineering. 2. Machinery. 3. Computer-aided engineering.
I. Fletcher, Adrian.
TJ145 .G56 2022
621--dc23

For information on all Willford Press publications
visit our website at www.willfordpress.com

Table of Contents

Preface

This book is a culmination of my many years of practice in this field. I attribute the success of this book to my support group. I would like to thank my parents who have showered me with unconditional love and support and my peers and professors for their constant guidance.

Mechanical engineering is the branch of engineering which involves the design, production and operation of machinery. It includes the application of various other disciplines such as engineering, physics, materials science and engineering mathematics to manufacture, analyze and maintain mechanical systems. It requires in-depth knowledge of areas such as thermodynamics, mechanics, structural analysis, dynamics and electricity. Mechanical engineering utilizes various tools like computer-aided design, product life cycle management and computer-aided manufacturing. This branch of engineering is used to design and analyze industrial equipment, manufacturing plants, heating and cooling systems, transport systems, robotics, aircraft, watercraft, and weapons. The topics included in this book on mechanical engineering are of utmost significance and bound to provide incredible insights to readers. Some of the diverse topics covered herein address the varied branches that fall under this category. This textbook will serve as a valuable source of reference for those interested in this field.

The details of chapters are provided below for a progressive learning:

Chapter - Mechanical Engineering: An Introduction

A machine is a mechanical structure which performs an intended function by using power for controlling movement and applying forces. The branch of engineering which seeks to design, manufacture, analyze and maintain machines is known as mechanical engineering. This chapter has been carefully written to provide an easy understanding of the varied facets of mechanical engineering and its importance.

Chapter - Branches of Mechanical Engineering

There are a number of different branches within mechanical engineering such as thermal engineering, fluid engineering, manufacturing engineering and machine design. Thermal engineering is a branch of engineering which focuses on the transfer and movement of heat energy. The topics elaborated in this chapter will help in gaining a better perspective about these branches of mechanical engineering.

Chapter - Fundamental Concepts of Mechanical Engineering

Mechanics and thermodynamics are two important concepts within mechanical engineering. Knowledge about structural analysis is also vital to this branch of engineering. The chapter closely examines these key concepts of mechanical engineering to provide an extensive understanding of the subject.

Chapter - Mechanical Vibrations

Mechanical vibration refers to the measurement of a periodic process of oscillations in relation to an equilibrium point. Some of the important areas of study within this field are vibration theory, string vibration and forced vibrations. All these diverse concepts of mechanical vibrations have been carefully analyzed in this chapter.

Chapter – Mechanics of Materials

Mechanics of materials is a field of study which deals with the behavior of solid objects as they are subjected to stresses and strains. Stress–strain curve, yield strength and compressive strength are some of the important concepts which are dealt with in this field. Mechanics of materials is a vast area that branches out into these significant sub-disciplines which have been thoroughly discussed in this chapter.

Chapter – Tools and Methods in Mechanical Engineering

There are a wide variety of tools used within mechanical engineering such as computer-aided design, computer -aided manufacturing and computational fluid dynamics. The use of computers to assist in creating, modifying, analyzing and optimizing a design is known as computer-aided design. The diverse applications of these tools in the field of mechanical engineering have been thoroughly discussed in this chapter.

Adrian Fletcher

1

Mechanical Engineering: An Introduction

A machine is a mechanical structure which performs an intended function by using power for controlling movement and applying forces. The branch of engineering which seeks to design, manufacture, analyze and maintain machines is known as mechanical engineering. This chapter has been carefully written to provide an easy understanding of the varied facets of mechanical engineering and its importance.

MACHINE

The scientific definition of a machine is any device that transmits or modifies energy. In common usage, the meaning is restricted to devices having rigid moving parts that perform or assist in performing some work. Machines normally require some energy source ("input") and always accomplish some sort of work ("output"). Devices with no rigid moving parts are commonly considered tools, or simply devices, not machines.

People have used mechanisms to amplify their abilities since before written records were available. Generally these devices decrease the amount of force required to do a given amount of work, alter the direction of the force, or transform one form of motion or energy into another.

Wind turbines

Modern power tools, automated machine tools, and human-operated power machinery are tools that are also machines. Machines used to transform heat or other energy into mechanical energy are known as engines.

Hydraulics devices may also be used to support industrial applications, although devices entirely lacking rigid moving parts are not commonly considered machines. Hydraulics are widely used in heavy equipment industries, automobile industries, marine industries, aeronautical industries, construction equipment industries, and earthmoving equipment industries.

Flint hand axe found in Winchester

Perhaps the first example of a human made device designed to manage power is the hand axe, made by chipping flint to form a wedge. A wedge is a simple machine that transforms lateral force and movement of the tool into a transverse splitting force and movement of the workpiece.

The idea of a simple machine originated with the Greek philosopher Archimedes around the third century B.C.E., who studied the Archimedean simple machines: lever, pulley, and screw. However the Greeks' understanding was limited to statics (the balance of forces) and did not include dynamics (the tradeoff between force and distance) or the concept of work.

During the Renaissance the dynamics of the Mechanical Powers, as the simple machines were called, began to be studied from the standpoint of how much useful work they could perform, leading eventually to the new concept of mechanical work. In 1586 Flemish engineer Simon Stevin derived the mechanical advantage of the inclined plane, and it was included with the other simple machines. The complete dynamic theory of simple machines was worked out by Italian scientist Galileo Galilei in 1600 in Le Meccaniche ("On Mechanics"). He was the first to understand that simple machines do not create energy, they merely transform it.

The classic rules of sliding friction in machines were discovered by Leonardo da Vinci, but remained unpublished in his notebooks. They were rediscovered by Guillaume Amontons and were further developed by Charles-Augustin de Coulomb.

Impact

Industrial Revolution

The Industrial Revolution was a period from 1750 to 1850 where changes in agriculture, manufacturing, mining, transportation, and technology had a profound effect on the social, economic, and cultural conditions of the times. It began in the United Kingdom, then subsequently spread throughout Western Europe, North America, Japan, and eventually the rest of the world.

Starting in the later part of the eighteenth century, there began a transition in parts of Great Britain's previously manual labor and draft-animal–based economy towards machine-based manufacturing. It started with the mechanization of the textile industries, the development of iron-making techniques and the increased use of refined coal.

Mechanization and Automation

Mechanization is providing human operators with machinery that assists them with the muscular requirements of work or displaces muscular work. In some fields, mechanization includes the use of hand tools. In modern usage, such as in engineering or economics, mechanization implies machinery more complex than hand tools and would not include simple devices such as an un-geared horse or donkey mill. Devices that cause speed changes or changes to or from reciprocating to rotary motion, using means such as gears, pulleys or sheaves and belts, shafts, cams and cranks, usually are considered machines. After electrification, when most small machinery was no longer hand powered, mechanization was synonymous with motorized machines.

Automation is the use of control systems and information technologies to reduce the need for human work in the production of goods and services. In the scope of industrialization, automation is a step beyond mechanization. Whereas mechanization provides human operators with machinery to assist them with the muscular requirements of work, automation greatly decreases the need for human sensory and mental requirements as well. Automation plays an increasingly important role in the world economy and in daily experience.

Automata

An automaton (plural: automata or automatons) is a self-operating machine. The word is sometimes used to describe a robot, more specifically an autonomous robot.

Types

The mechanical advantage of a simple machine is the ratio between the force it exerts on the load and the input force applied. This does not entirely describe the machine's performance, as force is required to overcome friction as well. The mechanical efficiency of a machine is the ratio of the actual mechanical advantage (AMA) to the ideal mechanical advantage (IMA). Functioning physical machines are always less than 100 percent efficient.

Mechanical

The word mechanical refers to the work that has been produced by machines or the machinery. It mostly relates to the machinery tools and the mechanical applications of science. Some of its synonyms are automatic and mechanic.

Simple Machines

The idea that a machine can be broken down into simple movable elements led Archimedes to define the lever, pulley and screw as simple machines. By the time of the Renaissance this list increased to include the wheel and axle, wedge and inclined plane.

Engines

An engine or motor is a machine designed to convert energy into useful mechanical motion. Heat engines, including internal combustion engines and external combustion engines (such as steam engines) burn a fuel to create heat, which is then used to create motion. Electric motors convert electrical energy into mechanical motion, pneumatic motors use compressed air and others, such as wind-up toys use elastic energy. In biological systems, molecular motors like myosins in muscles use chemical energy to create motion.

Electrical

Electrical means operating by or producing electricity, relating to or concerned with electricity. In other words, it means using, providing, producing, transmitting or operated by electricity.

Electrical Machine

An electrical machine is the generic name for a device that converts mechanical energy to electrical energy, converts electrical energy to mechanical energy, or changes alternating current from one voltage level to a different voltage level.

Electronic Machine

Electronics is the branch of physics, engineering and technology dealing with electrical circuits that involve active electrical components such as vacuum tubes, transistors, diodes and integrated circuits, and associated passive interconnection technologies. The nonlinear behavior of active components and their ability to control electron flows makes amplification of weak signals possible and is usually applied to information and signal processing. Similarly, the ability of electronic devices to act as switches makes digital information processing possible. Interconnection technologies such as circuit boards, electronic packaging technology, and other varied forms of communication infrastructure complete circuit functionality and transform the mixed components into a working system.

Computing Machines

Computers are machines to process information, often in the form of numbers. Charles Babbage designed various machines to tabulate logarithms and other functions in 1837. His Difference engine can be considered an advanced mechanical calculator and his Analytical Engine a forerunner of the modern computer, though none were built in Babbage's lifetime.

Modern computers are electronic ones. They use electric charge, current or magnetization to store and manipulate information. Computer architecture deals with detailed design of computers. There are also simplified models of computers, like State machine and Turing machine.

Molecular Machines

Study of the molecules and proteins that are the basis of biological functions has led to the concept of a molecular machine. For example, current models of the operation of the kinesin molecule that transports vesicles inside the cell as well as the myosin molecule that operates against actin to cause muscle contraction; these molecules control movement in response to chemical stimuli.

Researchers in nano-technology are working to construct molecules that perform movement in response to a specific stimulus. In contrast to molecules such as kinesin and myosin, these nano-machines or molecular machines are constructions like traditional machines that are designed to perform in a task.

A water-powered mine hoist used for raising ore. This woodblock is from De re metallica by Georg Bauer an early mining textbook that contains numerous drawings and descriptions of mining equipment.

Types of machines and related components		
Classification		Machine(s)
Simple machines		Inclined plane, Wheel and axle, Lever, Pulley, Wedge, Screw
Mechanical components		Axle, Bearings, Belts, Bucket, Fastener, Gear, Key, Link chains, Rack and pinion, Roller chains, Rope, Seals, Spring, Wheel
Clock		Atomic clock, Watch, Pendulum clock, Quartz clock
Compressors and Pumps		Archimedes' screw, Eductor-jet pump, Hydraulic ram, Pump, Trompe, Vacuum pump
Heat engines	External combustion engines	Steam engine, Stirling engine
	Internal combustion engines	Reciprocating engine, Gas turbine
Heat pumps		Absorption refrigerator, Thermoelectric refrigerator, Regenerative cooling
Linkages		Pantograph, Cam, Peaucellier-Lipkin
Turbine		Gas turbine, Jet engine, Steam turbine, Water turbine, Wind generator, Windmill
Aerofoil		Sail, Wing, Rudder, Flap, Propeller
Information technology		Computer, Calculator, Telecommunications networks
Electricity		Vacuum tube, Transistor, Diode, Resistor, Capacitor, Inductor, Memristor, Semiconductor
Robots		Actuator, Servo, Servomechanism, Stepper motor
Miscellaneous		Vending machine, Wind tunnel, Check weighing machines, Riveting machines

Machine Elements

Machines are assembled from standardized types of components. These elements consist of mechanisms that control movement in various ways such as gear trains, transistor switches, belt or chain drives, linkages, cam and follower systems, brakes and clutches, and structural components such as frame members and fasteners.

Modern machines include sensors, actuators and computer controllers. The shape, texture and color of covers provide a styling and operational interface between the mechanical components of a machine and its users.

Mechanisms

Assemblies within a machine that control movement are often called "mechanisms." Mechanisms are generally classified as gears and gear trains, cam and follower mechanisms, and linkages, though there are other special mechanisms such as clamping linkages, indexing mechanisms and friction devices such as brakes and clutches.

Controllers

Controllers combine sensors, logic, and actuators to maintain the performance of components of a machine. Perhaps the best known is the flyball governor for a steam engine. Examples of these devices range from a thermostat that as temperature rises opens a valve to cooling water to speed controllers such the cruise control system in an automobile. The programmable logic controller replaced relays and specialized control mechanisms with a programmable computer. Servo motors that accurately position a shaft in response to an electrical command are the actuators that make robotic systems possible.

MECHANICAL ENGINEERING

Technically, mechanical engineering is the application of the principles and problem-solving techniques of engineering from design to manufacturing to the marketplace for any object. Mechanical engineers analyze their work using the principles of motion, energy, and force—ensuring that designs function safely, efficiently, and reliably, all at a competitive cost.

Mechanical engineers make a difference. That's because mechanical engineering careers center on creating technologies to meet human needs. Virtually every product or service in modern life has probably been touched in some way by a mechanical engineer to help humankind.

This includes solving today's problems and creating future solutions in health care, energy, transportation, world hunger, space exploration, climate change, and more.

Being ingrained in many challenges and innovations across many fields means a mechanical engineering education is versatile. To meet this broad demand, mechanical engineers may design a component, a machine, a system, or a process. This ranges from the macro to the micro, from the largest systems like cars and satellites to the smallest components like sensors and switches.

Anything that needs to be manufactured—indeed, anything with moving parts—needs the expertise of a mechanical engineer.

Mechanical engineering combines creativity, knowledge and analytical tools to complete the difficult task of shaping an idea into reality.

This transformation happens at the personal scale, affecting human lives on a level we can reach out and touch like robotic prostheses. It happens on the local scale, affecting people in community-level spaces, like with agile interconnected microgrids. And it happens on bigger scales, like with advanced power systems, through engineering that operates nationwide or across the globe.

Mechanical engineers have an enormous range of opportunity and their education mirrors this breadth of subjects. Students concentrate on one area while strengthening analytical and problem-solving skills applicable to any engineering situation.

Disciplines within mechanical engineering include but are not limited to:

- Acoustics
- Aerospace
- Automation
- Automotive
- Autonomous Systems
- Biotechnology
- Composites
- Computer Aided Design (CAD)
- Control Systems
- Cyber security
- Design
- Energy
- Ergonomics
- Human health
- Manufacturing and additive manufacturing
- Mechanics
- Nanotechnology
- Production planning
- Robotics
- Structural analysis

Technology itself has also shaped how mechanical engineers work and the suite of tools has grown quite powerful in recent decades. Computer-aided engineering (CAE) is an umbrella term that covers everything from typical CAD techniques to computer-aided manufacturing to computer-aided engineering, involving finite element analysis (FEA) and computational fluid dynamics (CFD). These tools and others have further broadened the horizons of mechanical engineering.

Society depends on mechanical engineering. The need for this expertise is great in so many fields, and as such, there is no real limit for the freshly minted mechanical engineer. Jobs are always in demand, particularly in the automotive, aerospace, electronics, biotechnology, and energy industries.

Here are a handful of mechanical engineering fields.

In statics, research focuses on how forces are transmitted to and throughout a structure. Once a system is in motion, mechanical engineers look at dynamics, or what velocities, accelerations and resulting forces come into play. Kinematics then examines how a mechanism behaves as it moves through its range of motion.

Materials science delves into determining the best materials for different applications. A part of that is materials strength—testing support loads, stiffness, brittleness and other properties—which is essential for many construction, automobile, and medical materials.

How energy gets converted into useful power is the heart of thermodynamics, as well as determining what energy is lost in the process. One specific kind of energy, heat transfer, is crucial in many applications and requires gathering and analyzing temperature data and distributions.

Fluid mechanics, which also has a variety of applications, looks at many properties including pressure drops from fluid flow and aerodynamic drag forces.

Manufacturing is an important step in mechanical engineering. Within the field, researchers investigate the best processes to make manufacturing more efficient. Laboratory methods focus on improving how to measure both thermal and mechanical engineering products and processes. Likewise, machine design develops equipment-scale processes while electrical engineering focuses on circuitry. All this equipment produces vibrations, another field of mechanical engineering, in which researchers study how to predict and control vibrations.

Engineering economics makes mechanical designs relevant and usable in the real world by estimating manufacturing and life cycle costs of materials, designs, and other engineered products.

What Skills do Mechanical Engineers Need?

The essence of engineering is problem solving. With this at its core, mechanical engineering also requires applied creativity—a hands on understanding of the work involved—along with strong interpersonal skills like networking, leadership, and conflict management. Creating a product is only part of the equation; knowing how to work with people, ideas, data, and economics fully makes a mechanical engineer.

What Tasks do Mechanical Engineers do?

Careers in mechanical engineering call for a variety of tasks.

- Conceptual design
- Analysis
- Presentations and report writing
- Multidisciplinary teamwork
- Concurrent engineering
- Benchmarking the competition
- Project management
- Prototyping
- Testing
- Measurements
- Data Interpretation
- Developmental design
- Research
- Analysis (FEA and CFD)
- Working with suppliers
- Sales
- Consulting
- Customer service

IMPORTANCE OF MECHANICAL ENGINEERING

Mechanical engineers generally deal with the relations among forces, work or energy, and power in designing systems to improve the human environment. They may work to extract oil from deep within the earth or send a spacecraft to the moon. The products of their efforts may be automobiles or jet aircraft, nuclear power plants or air conditioning systems, large industrial machinery or household can openers. They are involved in programs to better utilize natural resources of energy and materials as well as to lessen the impact of technology on the environment.

Mechanical engineers, while strongly oriented towards science, are not scientists. Science is a search for knowledge. The science of mathematics extends abstract knowledge. The science of physics extends organized knowledge of the physical world. In each of these, consideration can be limited to a carefully isolated aspect of reality. The mechanical engineer must deal with reality in all its aspects. He or she must not only be competent to use the most classical and the most modern parts of science, but also must be able to devise and make a product which will be used by people. Moreover, the engineer must assume professional responsibility insofar as the safety and well-being of society are affected by those products.

A program in Mechanical Engineering will be a most stimulating and rewarding undergraduate experience for the great majority of students entering this field. Such a program is established by an educational environment created by men and women in contact with the world of people and industry. Engineering education is being called upon to produce graduates well-versed in rapidly advancing science, and who can lead industry and the public into the new world which engineering will make possible

Engineers will often discover in science, through their own research and invention or through the findings of scientists, those things which can be put to human use. In any engineering achievement, a new or better product is the objective; and all means available to the intellect of man will be employed to reach that objective. Science and its application remain a part, but only a part, of any great engineering advance. Young people who can respond to this kind of challenge are needed now, and they will be needed as never before in the years ahead.

2

Branches of Mechanical Engineering

There are a number of different branches within mechanical engineering such as thermal engineering, fluid engineering, manufacturing engineering and machine design. Thermal engineering is a branch of engineering which focuses on the transfer and movement of heat energy. The topics elaborated in this chapter will help in gaining a better perspective about these branches of mechanical engineering.

THERMAL ENGINEERING

Thermal engineering is a specialised sub-discipline of Mechanical Engineering that deals exclusively with heat energy and its transfer between not only different mediums, but also into other usable forms of energy. A Thermal engineer will be armed with the expertise to design systems and process to convert generated energy from various thermal sources into chemical, mechanical or electrical energy depending on the task at hand. Obviously, all thermal engineers are experts in all aspects of heat transfer.

Many process plants (basically somewhere where some raw material or resource is converted into something useful, e.g. power plants, oil refineries, plastic manufacturing plants, etc.) contain countless components and systems which have to be designed in terms of their heat transfer; it is particularly important to ensure that not too much heat is retained so the component or process is not disrupted. Conversely, some processes or systems are designed to use heat to their advantage and a thermal engineer must make sure enough heat is generated and used wisely (i.e. sustainably).

Thermal engineers must also know about the economics of the components and processes they design to make sure they not only provide an improvement over the existing solutions, but also don't lose the company money. Thermal engineers are not limited in areas of specialisation and can work in numerous fields. Below is only a brief example of areas a thermal engineer can work in:

- Heating, Ventilation and Air Conditioning (HVAC) systems in small and large-scale residential, commercial or industrial buildings.
- Renewable energy systems.
- Military and defence equipment.
- Electronics and electrical component and systems.

- Aerospace components.
- Boiler, heat exchanger, and pump design, amongst others.

Common industries that regularly employ thermal engineers includes power companies, the automotive industry and commercial construction. While thermal engineers will generally spend most of their time working in an office they are often required to travel to the site of their current project.

FLUID ENGINEERING

Fluid Mechanics is the branch of science that studies the behavior of fluids when they are in state of motion or rest. Whether the fluid is at rest or motion, it is subjected to different forces and different climatic conditions and it behaves in these conditions as per its physical properties. Fluid mechanics deals with three aspects of the fluid: static, kinematics, and dynamics aspects:

1. Fluid statics: The fluid which is in state of rest is called as static fluid and its study is called as fluid statics.
2. Fluid kinematics: The fluid which is in state of motion is called as moving fluid. The study of moving fluid without considering the effect of external pressures is called as fluid kinematics.
3. Fluid dynamics: The branch of science which studies the effect of all pressures including the external pressures on the moving fluid is called as fluid dynamics.

What is the necessity of stuying fluids as an aspect of engineering? Fluids are already an integral part of our day-to-day life. Engineering allows us to explore the potential of fluids for a number of new applications and various functions. Some of these include:

1. There are number of fluids that when burnt, produce lots of heat, which can be used for various applications. Examples of these fluids includes petrol and diesel for vehicles.
2. There are some fluids like oil that have a tendency to exert very high pressure or force. These fluids can be used for lifting various heavy loads. The fluids used in hydraulic machines and hydraulic lifters are an example.
3. Some fluids have excellent flow properties which can be used for the lubrication of various machines.
4. Fluids like water posses kinetic and potential energy, which is used for generation of electricity as in hydroelectric power plants.

Fluid mechanics helps us understand the behavior of fluid under various forces and at different atmospheric conditions, and to select the proper fluid for various applications.

This field is studied in detail within Civil Engineering and also to great extent in Mechanical Engineering and Chemical Engineering. It is in these branches of engineering where there is maximum use of the fluids.

Common Applications of Fluids

Hydroelectric Power Plants

In hydroelectric power plants, water is used to generate electricity on a large-scale basis. Water stored in the dam possesses potential energy, which is converted into the electrical energy in the power generation unit of the plant. Hydroelectric power plants are one of the major suppliers of power throughout the world. In some countries power requirements are fulfilled entirely by these plants.

Hydraulic Machines

Machines that operate on a fluid like water and oil are called hydraulic machines. The fluid as the capacity to lift heavy loads and exert extremely high pressures. Some hydraulic machines are used to perform various machining operations. In most of these machines, oil is used as the fluid. The oil is passed through the hydraulic motor which transfers large amounts of energy to the fluid. This high energy fluid enters the piston and cylinder arrangement where it can be used to lift heavy loads or apply large forces.

Automobiles

No automobile can run without fluid. Fluids perform three crucial operations in automobiles: generation of power, lubrication, and cooling of the engine. Petrol or diesel generates power on combustion in the engine. This is commonly referred to as fuel. Oil is used for the lubrication of the engine and the gearbox and also various other moving parts of the vehicle. In larger automobiles like cars, busses and trucks, water is used for cooling the engine.

Refrigerators and Air Conditioners

This is another important area where fluids play a crucial role. In refrigerators and air-conditioners, the fluids are known as refrigerants. The refrigerant absorbs the heat from whatever is being kept in the chiller or evaporator, which is at a low temperature, and delivers that heat to the atmosphere, which is at a high temperature. In air conditioners, the refrigerant absorbs room heat and throws it in to the atmosphere, thereby keeping the room cool. The entire operation of refrigerators and air-conditioners depends on the use of a refrigerant.

Thermal Power Plants

In thermal power plants, water is used as the working fluid. After getting heated in a boiler, water is converted into superheated steam which is passes through the blades of turbines, thus rotating them. The shaft of the turbine rotates in the generator, where electricity is produced. Thermal power plants are one of the major suppliers of power in various parts of the world, and water working as the fluid is their most important component.

Nuclear Power Plants

Water is again a crucial power plant component. Here it is both the working fluid and a coolant. In some nuclear power plants, heat produced within the nuclear reactor is used to directly heat water, which is converted into steam. This steam is passed through the turbines similar to thermal power plants, rotating turbine blades to generate power. This is an application of water as the working fluid.

In other nuclear power plants, the heat from nuclear reactors is not used to generate steam directly. Heat is first used to heat the water, which acts as the coolant. This coolant then transfers the heat to a secondary coolant or the working fluid, which is again water and it is passed through the turbine to generate electricity.

Fluids as a Renewable Energy Source

There are number of fluids that are being used as a renewable energy source. Air or wind is one of the most popular sources of renewable energy. Wind is used for generation of electricity on a small as well as large scale basis. Water is used in tidal power plants to generate electricity on a small scale basis. Ocean waves are used to rotate turbine blades within the power generation unit. Biodiesel, a type of the vegetable oil, is used as a fuel for vehicles along with traditional diesel.

Operating Various Instruments

Compressed air is used for the operation of various types of instruments and automatic valves. These valves can be activated and deactivated by applying the pressure of compressed air. The pneumatic tools which work on compressed air are used for various applications like grinding, screwing and unscrewing various machinery parts, etc.

Heat Engines

In previous heat engine designs, air was used as a fluid to generate power in automobiles. Earlier it was thought that the efficiency of an engine is dependent on the type of fluid used, but later it was shown by Sadi Carnot, that the efficiency of an engine is not dependent on the type of the fluid, but rather, the temperature of the fluid.

Fluids are used in a wide range of applications, often playing a vital role, without which, these applications will just cease to exist. The important thing to note is that most of the crucial applications of fluids are for generation of electricity or power. In hydroelectric power plants and automobiles, fluids are directly used to generate power or electricity. In thermal and nuclear power plants, fluids are indirectly used for generation of power, and still they are the dominant parts of these applications. It is not an overstatement to say that without fluids, the progress of the human race would stop.

MANUFACTURING ENGINEERING

Manufacturing Engineering is a branch of professional engineering that shares many common concepts and ideas with other fields of engineering such as mechanical,chemical, electrical and industrial engineering. Manufacturing engineering requires the ability to plan the practices of manufacturing; to research and to develop tools, processes, machines and equipment; and to integrate the facilities and systems for producing quality products with the optimum expenditure of capital.

The manufacturing or production engineer's primary focus is to turn raw material into an updated or new product in the most effective, efficient and economic way possible.

Manufacturing Engineering is based on core industrial engineering and mechanical engineering skills, adding important elements from mechatronics, commerce, economics and business management. This field also deals with the integration of different facilities and systems for producing quality products (with optimal expenditure) by applying the principles of physics and the results of manufacturing systems studies, such as the following:

- Craft or Guild.
- Putting-out system.
- British factory system.
- American system of manufacturing.
- Soviet collectivism in manufacturing.
- Mass production.
- Computer integrated manufacturing.
- Computer-aided technologies in manufacturing.
- Just in time manufacturing.
- Lean manufacturing.
- Flexible manufacturing.
- Mass customization.
- Agile manufacturing.
- Rapid manufacturing.
- Prefabrication.
- Ownership.
- Fabrication.
- Publication.

A set of six-axis robots used for welding.

Manufacturing engineers develop and create physical artifacts, production processes, and technology. It is a very broad area which includes the design and development of products. Manufacturing engineering is considered to be a subdiscipline of industrial engineering/systems engineering and has very strong overlaps with mechanical engineering. Manufacturing engineers' success or failure directly impacts the advancement of technology and the spread of innovation. This field of manufacturing engineering emerged from tool and die discipline in the early 20th century. It expanded greatly from the 1960s when industrialized countries introduced factories with:

1. Numerical control machine tools and automated systems of production.
2. Advanced statistical methods of quality control: These factories were pioneered by the American electrical engineer William Edwards Deming, who was initially ignored by his home country. The same methods of quality control later turned Japanese factories into world leaders in cost-effectiveness and production quality.
3. Industrial robots on the factory floor, introduced in the late 1970s: These computer-controlled welding arms and grippers could perform simple tasks such as attaching a car door quickly and flawlessly 24 hours a day. This cut costs and improved production speed.

The history of manufacturing engineering can be traced to factories in the mid 19th century USA and 18th century UK. Although large home production sites and workshops were established in China, ancient Rome and the Middle East, the Venice Arsenal provides one of the first examples of a factory in the modern sense of the word. Founded in 1104 in the Republic of Venice several hundred years before the Industrial Revolution, this factory mass-produced ships on assembly lines using manufactured parts. The Venice Arsenal apparently produced nearly one ship every day and, at its height, employed 16,000 people.

Many historians regard Matthew Boulton's Soho Manufactory (established in 1761 in Birmingham) as the first modern factory. Similar claims can be made for John Lombe's silk mill in Derby, or Richard Arkwright's Cromford Mill. The Cromford Mill was purpose-built to accommodate the equipment it held and to take the material through the various manufacturing processes.

Ford assembly line

One historian, Jack Weatherford, contends that the first factory was in Potosí. The Potosi factory took advantage of the abundant silver that was mined nearby and processed silver ingot slugs into coins.

British colonies in the 19th century built factories simply as buildings where a large number of workers gathered to perform hand labor, usually in textile production. This proved more efficient for the administration and distribution of materials to individual workers than earlier methods of manufacturing, such as cottage industries or the putting-out system.

Cotton mills used inventions such as the steam engine and the power loom to pioneer the industrial factories of the 19th century, where precision machine tools and replaceable parts allowed greater efficiency and less waste. This experience formed the basis for the later studies of manufacturing engineering. Between 1820 and 1850, non-mechanized factories supplanted traditional artisan shops as the predominant form of manufacturing institution.

Henry Ford further revolutionized the factory concept and thus manufacturing engineering in the early 20th century with the innovation of mass production. Highly specialized workers situated alongside a series of rolling ramps would build up a product such as (in Ford's case) an automobile. This concept dramatically decreased production costs for virtually all manufactured goods and brought about the age of consumerism.

Modern Developments

Modern manufacturing engineering studies include all intermediate processes required for the production and integration of a product's components.

Some industries, such as semiconductor and steel manufacturers use the term "fabrication" for these processes.

KUKA industrial robots being used at a bakery for food production.

Automation is used in different processes of manufacturing such as machining and welding. Automated manufacturing refers to the application of automation to produce goods in a factory. The main advantages of automated manufacturing for the manufacturing process are realized with effective implementation of automation and include: higher consistency and quality, reduction of lead times, simplification of production, reduced handling, improved work flow, and improved worker morale.

Robotics is the application of mechatronics and automation to create robots, which are often used in manufacturing to perform tasks that are dangerous, unpleasant, or repetitive. These robots may be of any shape and size, but all are preprogrammed and interact physically with the world. To create a robot, an engineer typically employs kinematics (to determine the robot's range of motion) and mechanics (to determine the stresses within the robot). Robots are used extensively in manufacturing engineering.

Robots allow businesses to save money on labor, perform tasks that are either too dangerous or too precise for humans to perform economically, and to ensure better quality. Many companies employ assembly lines of robots, and some factories are so robotized that they can run by themselves. Outside the factory, robots have been employed in bomb disposal, space exploration, and many other fields. Robots are also sold for various residential applications.

Modern Tools

Many manufacturing companies, especially those in industrialized nations, have begun to incorporate computer-aided engineering (CAE) programs into their existing design and analysis processes, including 2D and 3D solid modeling computer-aided design (CAD). This method has many benefits, including easier and more exhaustive visualization of products, the ability to create virtual assemblies of parts, and ease of use in designing mating interfaces and tolerances.

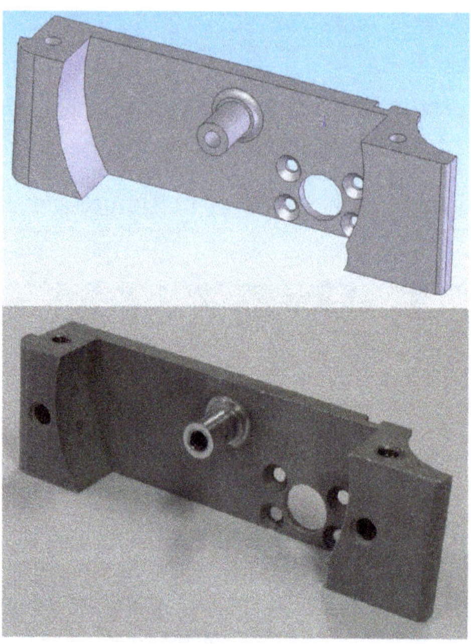

CAD model and CNC machined part.

Other CAE programs commonly used by product manufacturers include product life cycle management (PLM) tools and analysis tools used to perform complex simulations. Analysis tools may be used to predict product response to expected loads, including fatigue life and manufacturability. These tools include finite element analysis (FEA), computational fluid dynamics (CFD), and computer-aided manufacturing (CAM).

Using CAE programs, a mechanical design team can quickly and cheaply iterate the design process to develop a product that better meets cost, performance, and other constraints. No physical prototype need be created until the design nears completion, allowing hundreds or thousands of designs to be evaluated, instead of relatively few. In addition, CAE analysis programs can model complicated physical phenomena which cannot be solved by hand, such as viscoelasticity, complex contact between mating parts, or non-Newtonian flows.

Just as manufacturing engineering is linked with other disciplines, such as mechatronics, multidisciplinary design optimization (MDO) is also being used with other CAE programs to automate and improve the iterative design process. MDO tools wrap around existing CAE processes, allowing product evaluation to continue even after the analyst goes home for the day. They also utilize sophisticated optimization algorithms to more intelligently explore possible designs, often finding better, innovative solutions to difficult multidisciplinary design problems.

Manufacturing Engineering around the World

Manufacturing engineering is an extremely important discipline worldwide. It goes by different names in different countries. In the United States and the continental European Union it is commonly known as Industrial Engineering and in the United Kingdom and Australia it is called Manufacturing Engineering.

Subdisciplines

Mechanics

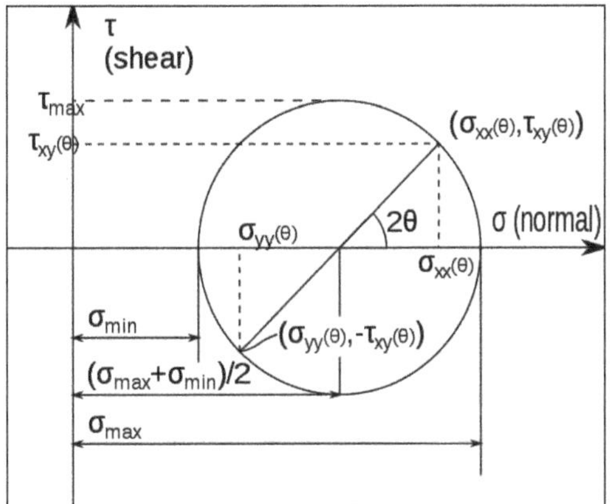

Mohr's circle, a common tool to study stresses in a mechanical element.

Mechanics, in the most general sense, is the study of forces and their effects on matter. Typically, engineering mechanics is used to analyze and predict the acceleration and deformation (both elastic and plastic) of objects under known forces (also called loads) or stresses. Subdisciplines of mechanics include:

- Statics, the study of non-moving bodies under known loads.
- Dynamics (or kinetics), the study of how forces affect moving bodies.
- Mechanics of materials, the study of how different materials deform under various types of stress.
- Fluid mechanics, the study of how fluids react to forces.
- Continuum mechanics, a method of applying mechanics that assumes that objects are continuous (rather than discrete).

If the engineering project were to design a vehicle, statics might be employed to design the frame of the vehicle in order to evaluate where the stresses will be most intense. Dynamics might be used when designing the car's engine to evaluate the forces in the pistons and cams as the engine cycles. Mechanics of materials might be used to choose appropriate materials for the manufacture of the frame and engine. Fluid mechanics might be used to design a ventilation system for the vehicle or to design the intake system for the engine.

Kinematics

Kinematics is the study of the motion of bodies (objects) and systems (groups of objects), while ignoring the forces that cause the motion. The movement of a crane and the oscillations of a piston in an engine are both simple kinematic systems. The crane is a type of open kinematic chain, while the piston is part of a closed four-bar linkage. Engineers typically use kinematics in the design and analysis of mechanisms. Kinematics can be used to find the possible range of motion for a given mechanism, or, working in reverse, can be used to design a mechanism that has a desired range of motion.

Drafting

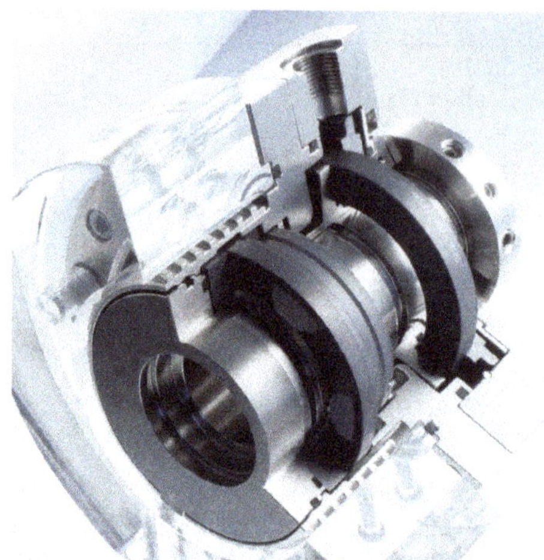

A CAD model of a mechanical double seal.

Drafting or technical drawing is the means by which manufacturers create instructions for manufacturing parts. A technical drawing can be a computer model or hand-drawn schematic showing all the dimensions necessary to manufacture a part, as well as assembly notes, a list of required materials, and other pertinent information. A U.S engineer or skilled worker who creates technical drawings may be referred to as a drafter or draftsman. Drafting has historically been a two-dimensional process, but computer-aided design (CAD) programs now allow the designer to create in three dimensions.

Instructions for manufacturing a part must be fed to the necessary machinery, either manually, through programmed instructions, or through the use of a computer-aided manufacturing (CAM) or combined CAD/CAM program. Optionally, an engineer may also manually manufacture a part using the technical drawings, but this is becoming an increasing rarity with the advent of computer numerically controlled (CNC) manufacturing. Engineers primarily manufacture parts manually in the areas of applied spray coatings, finishes, and other processes that cannot economically or practically be done by a machine.

Drafting is used in nearly every subdiscipline of mechanical and manufacturing engineering, and by many other branches of engineering and architecture. Three-dimensional models created using CAD software are also commonly used in finite element analysis (FEA) and computational fluid dynamics (CFD).

Machine Tools and Metal Fabrication

Machine tools employ some sort of tool that does the cutting or shaping. All machine tools have some means of constraining the workpiece and provide a guided movement of the parts of the machine. Metal fabrication is the building of metal structures by cutting, bending, and assembling processes.

Computer Integrated Manufacturing

Computer-integrated manufacturing (CIM) is the manufacturing approach of using computers to control the entire production process. Computer-integrated manufacturing is used in automotive, aviation, space, and ship building industries.

Mechatronics

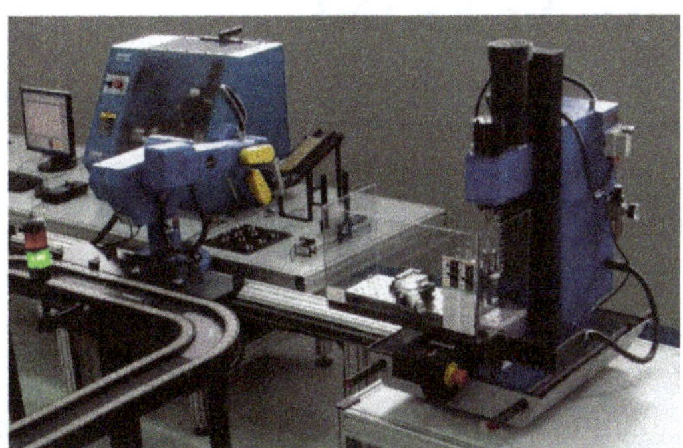

Training FMS with learning robot SCORBOT-ER 4u, workbench CNC mill and CNC lathe.

Mechatronics is an engineering discipline that deals with the convergence of electrical, mechanical and manufacturing systems. Such combined systems are known as electromechanical systems and are widespread. Examples include automated manufacturing systems, heating, ventilation and air-conditioning systems, and various aircraft and automobile subsystems.

The term mechatronics is typically used to refer to macroscopic systems, but futurists have predicted the emergence of very small electromechanical devices. Already such small devices, known as Microelectromechanical systems (MEMS), are used in automobiles to initiate the deployment of airbags, in digital projectors to create sharper images, and in inkjet printers to create nozzles for high-definition printing. In future it is hoped that such devices will be used in tiny implantable medical devices and to improve optical communication.

Textile Engineering

Textile engineering courses deal with the application of scientific and engineering principles to the design and control of all aspects of fiber, textile, and apparel processes, products, and machinery. These include natural and man-made materials, interaction of materials with machines, safety and health, energy conservation, and waste and pollution control. Additionally, students are given experience in plant design and layout, machine and wet process design and improvement, and designing and creating textile products. Throughout the textile engineering curriculum, students take classes from other engineering and disciplines including: mechanical, chemical, materials and industrial engineering.

Advanced Composite Materials

Advanced composite materials (engineering) (ACMs) are also known as advanced polymer matrix composites. These are generally characterized or determined by unusually high strength fibres with unusually high stiffness, or modulus of elasticity characteristics, compared to other materials, while bound together by weaker matrices. Advanced composite materials have broad, proven applications, in the aircraft, aerospace, and sports equipment sectors. Even more specifically ACMs are very attractive for aircraft and aerospace structural parts. Manufacturing ACMs is a multibillion-dollar industry worldwide. Composite products range from skateboards to components of the space shuttle. The industry can be generally divided into two basic segments, industrial composites and advanced composites.

AUTOMOTIVE ENGINEERING

Automotive engineering, along with aerospace engineering and naval architecture, is a branch of vehicle engineering, incorporating elements of mechanical, electrical, electronic, software, and safety engineering as applied to the design, manufacture and operation of motorcycles, automobiles, and trucks and their respective engineering subsystems. It also includes modification of vehicles. Manufacturing domain deals with the creation and assembling the whole parts of automobiles is also included in it. The automotive engineering field is research -intensive and involves direct application of mathematical models and formulas. The study of automotive engineering is to design, develop, fabricate, and test vehicles or vehicle components from the concept stage to production stage. Production, development, and manufacturing are the three major functions in this field.

Disciplines

Automobile Engineering

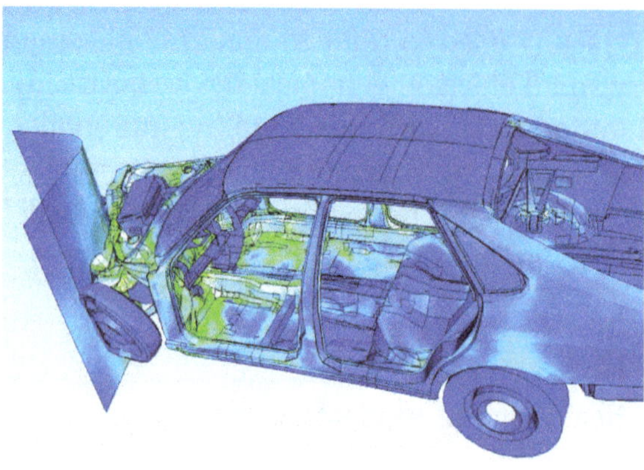

Visualization of how a car deforms in an asymmetrical crash using finite element analysis.

Automobile Engineering is a branch study of engineering which teaches manufacturing, designing, mechanical mechanisms as well operations of automobiles. It is an introduction to vehicle engineering which deals with motorcycles, cars, buses, trucks, etc. It includes branch study of mechanical, electronic, software and safety elements. Some of the engineering attributes and disciplines that are of importance to the automotive engineer and many of the other aspects are included in it:

Safety engineering: Safety engineering is the assessment of various crash scenarios and their impact on the vehicle occupants. These are tested against very stringent governmental regulations. Some of these requirements include: seat belt and air bag functionality testing, front and side impact testing, and tests of rollover resistance. Assessments are done with various methods and tools, including Computer crash simulation (typically finite element analysis), crash test dummy, and partial system sled and full vehicle crashes.

Fuel economy/emissions: Fuel economy is the measured fuel efficiency of the vehicle in miles per gallon or kilometers per liter. Emissions testing includes the measurement of vehicle emissions, including hydrocarbons, nitrogen oxides (NO_x), carbon monoxide (CO), carbon dioxide (CO_2), and evaporative emissions.

NVH engineering (noise, vibration, and harshness): NVH is the customer's feedback (both tactile felt and audible heard) from the vehicle. While sound can be interpreted as a rattle, squeal, or hot, a tactile response can be seat vibration or a buzz in the steering wheel. This feedback is generated by components either rubbing, vibrating, or rotating. NVH response can be classified in various ways: powertrain NVH, road noise, wind noise, component noise, and squeak and rattle. Note, there are both good and bad NVH qualities. The NVH engineer works to either eliminate bad NVH or change the "bad NVH" to good (i.e., exhaust tones).

Vehicle electronics: Automotive electronics is an increasingly important aspect of automotive engineering. Modern vehicles employ dozens of electronic systems. These systems are responsible for operational controls such as the throttle, brake and steering controls; as well as many comfort and convenience systems such as the HVAC, infotainment, and lighting systems. It would not be possible for automobiles to meet modern safety and fuel economy requirements without electronic controls.

Performance: Performance is a measurable and testable value of a vehicle's ability to perform in various conditions. Performance can be considered in a wide variety of tasks, but it's generally associated with how quickly a car can accelerate (e.g. standing start 1/4 mile elapsed time, 0–60 mph, etc.), its top speed, how short and quickly a car can come to a complete stop from a set speed (e.g. 70-0 mph), how much g-force a car can generate without losing grip, recorded lap times, cornering speed, brake fade, etc. Performance can also reflect the amount of control in inclement weather (snow, ice, rain).

Shift quality: Shift quality is the driver's perception of the vehicle to an automatic transmission shift event. This is influenced by the powertrain (engine, transmission), and the vehicle (driveline, suspension, engine and powertrain mounts, etc.). Shift feel is both a tactile (felt) and audible (heard) response of the vehicle. Shift quality is experienced as various events: Transmission shifts are felt as an upshift at acceleration (1–2), or a downshift maneuver in passing (4–2). Shift engagements of the vehicle are also evaluated, as in Park to Reverse, etc.

Durability/corrosion engineering: Durability and corrosion engineering is the evaluation testing of a vehicle for its useful life. Tests include mileage accumulation, severe driving conditions, and corrosive salt baths.

Drivability: Drivability is the vehicle's response to general driving conditions. Cold starts and stalls, RPM dips, idle response, launch hesitations and stumbles, and performance levels.

Cost: The cost of a vehicle program is typically split into the effect on the variable cost of the vehicle, and the up-front tooling and fixed costs associated with developing the vehicle. There are also costs associated with warranty reductions and marketing.

Program timing: To some extent programs are timed with respect to the market, and also to the production schedules of the assembly plants. Any new part in the design must support the development and manufacturing schedule of the model.

Assembly feasibility: It is easy to design a module that is hard to assemble, either resulting in damaged units or poor tolerances. The skilled product development engineer works with the assembly/ manufacturing engineers so that the resulting design is easy and cheap to make and assemble, as well as delivering appropriate functionality and appearance.

Quality management: Quality control is an important factor within the production process, as high quality is needed to meet customer requirements and to avoid expensive recall campaigns. The complexity of components involved in the production process requires a combination of different tools and techniques for quality control. Therefore, the International Automotive Task Force (IATF), a group of the world's leading manufacturers and trade organizations, developed the standard ISO/TS 16949. This standard defines the design, development, production, and when relevant, installation and service requirements. Furthermore, it combines the principles of ISO 9001 with aspects of various regional and national automotive standards such as AVSQ (Italy), EAQF (France), VDA6 (Germany) and QS-9000 (USA). In order to further minimize risks related to product failures and liability claims of automotive electric and electronic systems, the quality discipline functional safety according to ISO/IEC 17025 is applied.

Since the 1950s, the comprehensive business approach total quality management, TQM, helps to continuously improve the production process of automotive products and components. Some of the companies who have implemented TQM include Ford Motor Company, Motorola and Toyota Motor Company.

The Modern Automotive Product Engineering Process

Studies indicate that a substantial part of the modern vehicle's value comes from intelligent systems, and that these represent most of the current automotive innovation. To facilitate this, the modern automotive engineering process has to handle an increased use of mechatronics. Configuration and performance optimization, system integration, control, component, subsystem and system-level validation of the intelligent systems must become an intrinsic part of the standard vehicle engineering process, just as this is the case for the structural, vibro-acoustic and kinematic design. This requires a vehicle development process that is typically highly simulation-driven.

The V-Approach

One way to effectively deal with the inherent multi-physics and the control systems development that is involved when including intelligent systems, is to adopt the V-Model approach to systems development, as has been widely used in the automotive industry for twenty years or more. In this V-approach, system-level requirements are propagated down the V via subsystems to component design, and the system performance is validated at increasing integration levels. Engineering of mechatronic systems requires the application of two interconnected "V-cycles": one focusing on the multi-physics system engineering (like the mechanical and electrical components of an electrically powered steering system, including sensors and actuators); and the other focuses on the controls engineering, the control logic, the software and realization of the control hardware and embedded software.

Predictive Engineering Analytics

An alternative approach is called predictive engineering analytics, and takes the V-approach to the next level. It lets design continue after product delivery. That is important for development of built-in predictive functionality and for creating vehicles that can be optimized while being in use, even based on real use data. This approach is based on the creation of a Digital Twin, a replica of the real product that remains in-sync. Manufacturers try to achieve this by implementing a set of development tactics and tools. Critical is a strong alignment of 1D systems simulation, 3D CAE and physical testing to reach more realism in the simulation process. This is combined with intelligent reporting and data analytics for better insight in the vehicle use. By supporting this with a strong data management structure that spans the entire product lifecycle, they bridge the gap between design, manufacturing and product use.

MECHATRONICS

Mechatronics, which is also called mechatronic engineering, is a multidisciplinary branch of engineering that focuses on the engineering of both electrical and mechanical systems, and also includes a combination of robotics, electronics, computer, telecommunications, systems, control, and product engineering. As technology advances over time, various subfields of engineering have succeeded in both adapting and multiplying. The intention of mechatronics is to produce a design solution that unifies each of these various subfields. Originally, the field of mechatronics was intended to be nothing more than a combination of mechanics and electronics, hence the name being a portmanteau of mechanics and electronics; however, as the complexity of technical systems continued to evolve, the definition had been broadened to include more technical areas.

The word *mechatronics* originated in Japanese-English and was created by Tetsuro Mori, an engineer of Yaskawa Electric Corporation. The word *mechatronics* was registered as trademark by the company in Japan with the registration number of "46-32714" in 1971. However, afterward the company released the right of using the word to public, the word begun being used across the world. Nowadays, the word is translated into many languages and the word is considered as an essential term for industry.

French standard NF E 01-010 gives the following definition: "approach aiming at the synergistic integration of mechanics, electronics, control theory, and computer science within product design and manufacturing, in order to improve and/or optimize its functionality".

Many people treat *mechatronics* as a modern buzzword synonymous with robotics and electromechanical engineering.

Mechatronic system.

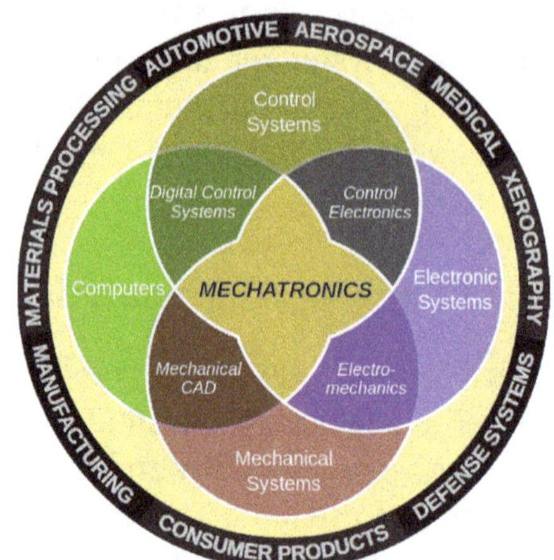

Aerial Euler diagram from RPI's website describes the fields that make up mechatronics.

A mechatronics engineer unites the principles of mechanics, electronics, and computing to generate a simpler, more economical and reliable system. The term "mechatronics" was coined by Tetsuro Mori, the senior engineer of the Japanese company Yaskawa in 1969. An industrial robot is a prime example of a mechatronics system; it includes aspects of electronics, mechanics, and computing to do its day-to-day jobs.

Engineering cybernetics deals with the question of control engineering of mechatronic systems. It is used to control or regulate such a system. Through collaboration, the mechatronic modules perform the production goals and inherit flexible and agile manufacturing properties in the production scheme. Modern production equipment consists of mechatronic modules that are integrated according to a control architecture. The most known architectures involve hierarchy, polyarchy, heterarchy, and hybrid. The methods for achieving a technical effect are described by control algorithms, which might or might not utilize formal methods in their design. Hybrid systems important to mechatronics include production systems, synergy drives, planetary exploration rovers, automotive subsystems such as anti-lock braking systems and spin-assist, and everyday equipment such as autofocus cameras, video, hard disks, and CD players.

Course Structure

Mechatronics students take courses in various fields:

- Mechanical engineering and materials science and engineering.
- Electronics engineering.
- Electrical engineering.
- Computer engineering (software and hardware engineering).
- Computer science.
- Systems engineering.

- Control engineering.
- Optical engineering.
- Telecommunications.

Applications

- Machine vision.
- Automation and robotics.
- Servo-mechanics.
- Sensing and control systems.
- Automotive engineering, automotive equipment in the design of subsystems such as anti-lock braking systems.
- Computer-machine controls, such as computer driven machines like CNC milling machines, CNC waterjets, and CNC plasma cutters.
- Expert systems.
- Industrial goods.
- Consumer products.
- Mechatronics systems.
- Medical mechatronics, medical imaging systems.
- Structural dynamic systems.
- Transportation and vehicular systems.
- Mechatronics as the new language of the automobile.
- Computer aided and integrated manufacturing systems.
- Computer-aided design.
- Engineering and manufacturing systems.
- Packaging.
- Microcontrollers/PLCs.

Physical Implementations

Mechanical modeling calls for modeling and simulating physical complex phenomena in the scope of a multi-scale and multi-physical approach. This implies to implement and to manage modeling and optimization methods and tools, which are integrated in a systemic approach. The specialty is aimed for students in mechanics who want to open their mind to systems engineering, and able to integrate different physics or technologies, as well as students in mechatronics who want to increase their knowledge in optimization and multidisciplinary simulation techniques. The speciality educates students in robust and/or optimized conception methods for structures or many technological systems, and to the main modeling and simulation tools used in R&D. Special courses are also proposed for original applications (multi-materials composites, innovating transducers and actuators, integrated systems) to prepare the students to the coming breakthrough in the domains covering the materials and the systems. For some mechatronic systems, the main issue is no longer how to implement a control system, but how to implement actuators. Within the mechatronic field, mainly two technologies are used to produce movement/motion.

Variant of the Field

An emerging variant of this field is biomechatronics, whose purpose is to integrate mechanical parts with a human being, usually in the form of removable gadgets such as an exoskeleton. This is the "real-life" version of cyberware.

Another variant that we can consider is Motion control for Advanced Mechatronics, which presently is recognized as a key technology in mechatronics. The robustness of motion control will be represented as a function of stiffness and a basis for practical realization. Target of motion is parameterized by control stiffness which could be variable according to the task reference. However, the system robustness of motion always requires very high stiffness in the controller.

Avionics is also considered a variant of mechatronics as it combines several fields such as electronics and telecom with Aerospace engineering.

Internet of Things

The Internet of things (IoT) is the inter-networking of physical devices, embedded with electronics, software, sensors, actuators, and network connectivity which enable these objects to collect and exchange data.

IoT and mechatronics are complementary. Many of the smart components associated with the Internet of Things will be essentially mechatronic. The development of the IoT is forcing mechatronics engineers, designers, practitioners and educators to research the ways in which mechatronic systems and components are perceived, designed and manufactured. This allows them to face up to new issues such as data security, machine ethics and the human-machine interface.

MACHINE DESIGN

Mechanical Design or Machine Design is one of the important branches of Engineering Design. To understand what exactly machine design or mechanical design is let us consider the example of the gear box of the car. The gear box transmits the motion and the power of the engine to the wheels of the vehicle. The gearbox comprises group of gears which are subjected to not only motion but also the load of the vehicle. For the gears to run at desired speeds and take desired loads it is important that they should be designed. During designing various calculations are performed considering desired speeds and loads and finally the gear of particular material and specific dimensions that can take all loads and that can be manufactured at least possible cost giving optimum performance is designed. In similar fashion all the components of the car, including engine, have to be designed so that they optimally meet all the functional requirements at lowest possible cost. This whole process of designing is called as machine design or mechanical design.

Machine Design or Mechanical Design can be defined as the process by which resources or energy is converted into useful mechanical forms, or the mechanisms so as to obtain useful output from the machines in the desired form as per the needs of the human beings. Machine design can lead to the formation of the entirely new machine or it can lead to up-gradation or improvement of the existing machine. For instance if the existing gearbox is too heavy or cannot sustain the actual loads, entirely new gearbox can be designed. But if the same gearbox has the potential to lift more loads, it can be upgraded by making certain important changes in its design.

References

- Lawrence J. Kamm (1996). Understanding Electro-Mechanical Engineering: An Introduction to Mechatronics. John Wiley & Sons. ISBN 978-0-7803-1031-5
- What-is-thermal-engineering: whatisengineering.com, Retrieved 11 May, 2019
- Van der Auweraer, Herman; Anthonis, Jan; De Bruyne, Stijn; Leuridan, Jan (July 2013). "Virtual engineering at work: the challenges for designing mechatronic products". Engineering with Computers. 29 (3): 389–408. doi:10.1007/s00366-012-0286-6
- Fluid-mechanics-hydraulics, what-is-fluid-mechanics-9704: brighthubengineering.com, Retrieved 12 June, 2019
- Mechanical and Mechatronics Engineering Department. "What is Mechatronics Engineering?". Prospective Student Information. University of Waterloo. Retrieved 30 May, 2011
- Cad-autocad-reviews-tips, what-is-mechanical-design-or-machine-design-9935: brighthubengineering.com, Retrieved 30 April, 2019

3

Fundamental Concepts of Mechanical Engineering

Mechanics and thermodynamics are two important concepts within mechanical engineering. Knowledge about structural analysis is also vital to this branch of engineering. The chapter closely examines these key concepts of mechanical engineering to provide an extensive understanding of the subject.

MECHANICS

Mechanics is the science concerned with the motion of bodies under the action of forces, including the special case in which a body remains at rest. Of first concern in the problem of motion are the forces that bodies exert on one another. This leads to the study of such topics as gravitation, electricity, and magnetism, according to the nature of the forces involved. Given the forces, one can seek the manner in which bodies move under the action of forces; this is the subject matter of mechanics proper.

Historically, mechanics was among the first of the exact sciences to be developed. Its internal beauty as a mathematical discipline and its early remarkable success in accounting in quantitative detail for the motions of the Moon, the Earth, and other planetary bodies had enormous influence on philosophical thought and provided impetus for the systematic development of science into the 20th century.

Mechanics may be divided into three branches: statics, which deals with forces acting on and in a body at rest; kinematics, which describes the possible motions of a body or system of bodies; and kinetics, which attempts to explain or predict the motion that will occur in a given situation. Alternatively, mechanics may be divided according to the kind of system studied. The simplest mechanical system is the particle, defined as a body so small that its shape and internal structure are of no consequence in the given problem. More complicated is the motion of a system of two or more particles that exert forces on one another and possibly undergo forces exerted by bodies outside of the system.

The principles of mechanics have been applied to three general realms of phenomena. The motions of such celestial bodies as stars, planets, and satellites can be predicted with great accuracy thousands of years before they occur. (The theory of relativity predicts some deviations from the motion according to classical, or Newtonian, mechanics; however, these are so small as to be observable only with very accurate techniques, except in problems involving all or a large portion of the detectable universe.) As the second realm, ordinary objects on Earth down to microscopic size (moving at speeds much lower than that of light) are properly described by classical mechanics without significant corrections. The engineer who designs bridges or aircraft may use the Newtonian laws of classical mechanics with confidence, even though the forces may be very complicated, and the calculations lack the beautiful simplicity of celestial mechanics. The third realm of phenomena comprises the behaviour of matter and electromagnetic radiation on the atomic and subatomic scale. Although there were some limited early successes in describing the behaviour of atoms in terms of classical mechanics, these phenomena are properly treated in quantum mechanics.

Classical mechanics deals with the motion of bodies under the influence of forces or with the equilibrium of bodies when all forces are balanced. The subject may be thought of as the elaboration and application of basic postulates first enunciated by Isaac Newton in his Philosophiae Naturalis Principia Mathematica, commonly known as the Principia. These postulates, called Newton's laws of motion, are set forth below. They may be used to predict with great precision a wide variety of phenomena ranging from the motion of individual particles to the interactions of highly complex systems.

In the framework of modern physics, classical mechanics can be understood to be an approximation arising out of the more profound laws of quantum mechanics and the theory of relativity. However, that view of the subject's place greatly undervalues its importance in forming the context, language, and intuition of modern science and scientists. Our present-day view of the world and man's place in it is firmly rooted in classical mechanics. Moreover, many ideas and results of classical mechanics survive and play an important part in the new physics.

The central concepts in classical mechanics are force, mass, and motion. Neither force nor mass is very clearly defined by Newton, and both have been the subject of much philosophical speculation since Newton. Both of them are best known by their effects. Mass is a measure of the tendency of a body to resist changes in its state of motion. Forces, on the other hand, accelerate bodies, which is to say, they change the state of motion of bodies to which they are applied. The interplay of these effects is the principal theme of classical mechanics.

Although Newton's laws focus attention on force and mass, three other quantities take on special importance because their total amount never changes. These three quantities are energy, (linear) momentum, and angular momentum. Any one of these can be shifted from one body or system of bodies to another. In addition, energy may change form while associated with a single system, appearing as kinetic energy, the energy of motion; potential energy, the energy of position; heat, or internal energy, associated with the random motions of the atoms or molecules composing any real body; or any combination of the three. Nevertheless, the total energy, momentum, and angular momentum in the universe never changes. This fact is expressed in physics by saying that energy, momentum, and angular momentum are conserved. These three conservation laws arise out of Newton's laws, but Newton himself did not express them. They had to be discovered later.

It is a remarkable fact that, although Newton's laws are no longer considered to be fundamental, nor even exactly correct, the three conservation laws derived from Newton's laws—the conservation of energy, momentum, and angular momentum—remain exactly true even in quantum mechanics and relativity. In fact, in modern physics, force is no longer a central concept, and mass is only one of a number of attributes of matter. Energy, momentum, and angular momentum, however, still firmly hold centre stage. The continuing importance of these ideas inherited from classical mechanics may help to explain why this subject retains such great importance in science today.

Fundamental Concepts

Units and Dimensions

Quantities have both dimensions, which are an expression of their fundamental nature, and units, which are chosen by convention to express magnitude or size. For example, a series of events have a certain duration in time. Time is the dimension of the duration. The duration might be expressed as 30 minutes or as half an hour. Minutes and hours are among the units in which time may be expressed. One can compare quantities of the same dimensions, even if they are expressed in different units (an hour is longer than a minute). Quantities of different dimensions cannot be compared with one another.

The fundamental dimensions used in mechanics are time, mass, and length. Symbolically, these are written as t, m, and l, respectively. The study of electromagnetism adds an additional fundamental dimension, electric charge, or q. Other quantities have dimensions compounded of these. For example, speed has the dimensions distance divided by time, which can be written as l/t, and volume has the dimensions distance cubed, or l^3. Some quantities, such as temperature, have units but are not compounded of fundamental dimensions.

There are also important dimensionless numbers in nature, such as the number $\pi = 3.14159....$ Dimensionless numbers may be constructed as ratios of quantities having the same dimension. Thus, the number π is the ratio of the circumference of a circle (a length) to its diameter (another length). Dimensionless numbers have the advantage that they are always the same, regardless of what set of units is being used.

Governments have traditionally been responsible for establishing and enforcing standard units for the sake of orderly commerce, navigation, science, and, of course, taxation. Today all such units are established by international treaty, revised every few years in light of scientific findings. The units used for most scientific measurements are those designated the International System of Units (Systѐme International d'Unités), or SI for short. They are based on the metric system, first adopted officially by France in 1795. Other units, such as those of the British engineering system, are still in use in some places, but these are now defined in terms of the SI units.

The fundamental unit of length is the metre. A metre used to be defined as the distance between two scratch marks on a metal bar kept in Paris, but it is now much more precisely defined as the distance that light travels in a certain time interval (1/299,792,458 of a second). By contrast, in the British system, units of length have a clear human bias: the foot, the inch (the first joint of the thumb), the yard (distance from nose to outstretched fingertip), and the mile (one thousand standard paces of a Roman legion). Each of these is today defined as some fraction or multiple of a metre (one yard is nearly equal to one metre). In the SI or the metric system, lengths are expressed as decimal fractions or multiples of a metre (a millimetre = one-thousandth of a metre; a centimetre = one-hundredth of a metre; a kilometre = one thousand metres).

Times longer than one second are expressed in the units seconds, minutes, hours, days, weeks, and years. Times shorter than one second are expressed as decimal fractions (a millisecond = one-thousandth of a second, a microsecond = one-millionth of a second, and so on). The fundamental unit of time (i.e., the definition of one second) is today based on the intrinsic properties of certain kinds of atoms (an excitation frequency of the isotope cesium-133).

Units of mass are also defined in a way that is technically sound, but in common usage they are the subject of some confusion because they are easily confused with units of weight, which is a different physical quantity. The weight of an object is the consequence of the Earth's gravity operating on its mass. Thus, the mass of a given object is the same everywhere, but its weight varies slightly if it is moved about the surface of the Earth, and it would change a great deal if it were moved to the surface of another planet. Also, weight and mass do not have the same dimensions (weight has the dimensions ml/t^2). The Constitution of the United States, which calls on the government to establish uniform "weights and measures," is oblivious to this distinction, as are merchants the world over, who measure the weight of bread or produce but sell it in units of kilograms, the SI unit of mass. (The kilogram is equal to 1,000 grams; 1 gram is the mass of 1 cubic centimetre of water—under appropriate conditions of temperature and pressure.)

Vectors

The equations of mechanics are typically written in terms of Cartesian coordinates. At a certain time t, the position of a particle may be specified by giving its coordinates x(t), y(t), and z(t) in a particular Cartesian frame of reference. However, a different observer of the same particle might choose a differently oriented set of mutually perpendicular axes, say, x′, y′, and z′. The motion of the particle is then described by the first observer in terms of the rate of change of x(t), y(t), and z(t), while the second observer would discuss the rates of change of x′(t), y′(t), and z′(t). That is, both observers see the same particle executing the same motion and obeying the same laws, but they describe the situation with different equations. This awkward situation may be avoided by means of a mathematical construction called a vector. Although vectors are mathematically simple and extremely useful in discussing mechanics, they were not developed in their modern form until late in the 19th century, when J. Willard Gibbs and Oliver Heaviside (of the United States and Britain, respectively) each applied vector analysis in order to help express the new laws of electromagnetism proposed by James Clerk Maxwell.

A vector is a quantity that has both magnitude and direction. It is typically represented symbolically by an arrow in the proper direction, whose length is proportional to the magnitude of the vector. Although a vector has magnitude and direction, it does not have position. A vector is not altered if it is displaced parallel to itself as long as its length is not changed.

By contrast to a vector, an ordinary quantity having magnitude but not direction is known as a scalar. In printed works vectors are often represented by boldface letters such as A or X, and scalars are represented by lightface letters, A or X. The magnitude of a vector, denoted $|A|$, is itself a scalar—i.e., $|A| = A$.

Because vectors are different from ordinary (i.e., scalar) quantities, all mathematical operations involving vectors must be carefully defined. There is no mathematical operation that corresponds to division by a vector.

If vector A is added to vector B, the result is another vector, C, written A + B = C. The operation is performed by displacing B so that it begins where A ends, as shown in figure. C is then the vector that starts where A begins and ends where B ends.

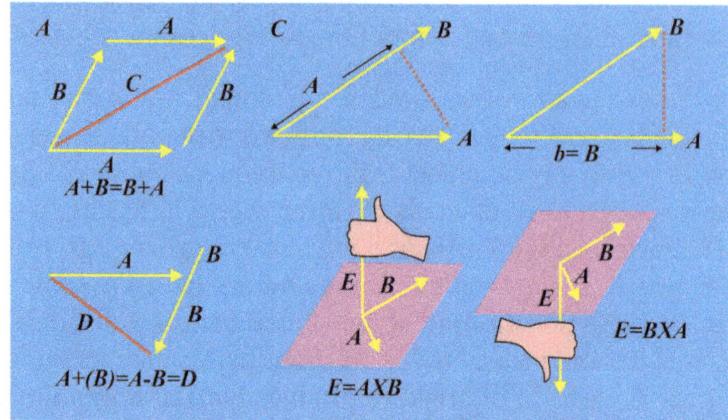

The vector sum C = A + B = B + A. (B) The vector difference A + (−B) = A − B = D. (C, left) A cos θ is the component of A along B and (right) B cos θ is the component of B along A. (D, left) The right-hand rule used to find the direction of E = A × B and (right) the right-hand rule used to find the direction of −E = B × A.

Vector addition is defined to have the (nontrivial) property A + B = B + A. There do exist quantities having magnitude and direction that do not obey this requirement. An example is finite rotations in space. Two finite rotations of a body about different axes do not necessarily result in the same orientation if performed in the opposite order.

Vector subtraction is defined by A − B = A + (−B), where the vector −B has the same magnitude as B but the opposition direction. The idea is illustrated in figure.

A vector may be multiplied by a scalar. Thus, for example, the vector 2A has the same direction as A but is twice as long. If the scalar has dimensions, the resulting vector still has the same direction as the original one, but the two cannot be compared in magnitude. For example, a particle moving with constant velocity v suffers a displacement s in time t given by s = vt. The vector v has been multiplied by the scalar t to give a new vector, s, which has the same direction as v but cannot be compared to v in magnitude (a displacement of one metre is neither bigger nor smaller than a velocity of one metre per second). This is a typical example of a phenomenon that might be represented by different equations in differently oriented Cartesian coordinate systems but that has a single vector equation (for all observers not moving with respect to one another).

The dot product (also known as the scalar product, or sometimes the inner product) is an operation that combines two vectors to form a scalar. The operation is written $A \cdot B$. If θ is the (smaller) angle between A and B, then the result of the operation is $A \cdot B = AB \cos \theta$. The dot product measures the extent to which two vectors are parallel. It may be thought of as multiplying the magnitude of one vector (either one) by the projection of the other upon it, as shown in figure. If the two vectors are perpendicular, the dot product is zero.

The cross product (also known as the vector product) combines two vectors to form another vector, perpendicular to the plane of the original vectors. The operation is written A × B. If θ is the (smaller) angle between A and B, then $|A \times B| = AB \sin \theta$. The direction of A × B is given by the right-hand rule: if the fingers of the right hand are made to rotate from A through θ to B, the thumb points in the direction of A × B, as shown in figure. The cross product is zero if the two vectors are parallel, and it is maximum in magnitude if they are perpendicular.

The derivative, or rate of change, of a vector is defined in perfect analogy to the derivative of a scalar: if the vector A changes with time t, then,

$$\frac{dA}{dt} = \lim_{\Delta t \to 0} \frac{A(t+\Delta t) - A(t)}{\Delta t}$$

Before going to the limit on the right-hand side of equation $\frac{dA}{dt} = \lim_{\Delta t \to 0} \frac{A(t+\Delta t) - A(t)}{\Delta t}$, the operations described are vector subtraction [A(t + Δt) – A(t)] and scalar multiplication (by 1/Δt). The result, dA/dt, is therefore itself a vector. Notice that, as shown in figure, the difference between two vectors, in this case A(t + Δt) – A(t), may be in quite a different direction than either of the vectors from which it is formed, here A(t + Δt) and A(t). As a result, dA/dt may be in a different direction than A(t).

Newton's Laws of Motion and Equilibrium

In his Principia, Newton reduced the basic principles of mechanics to three laws:

1. Every body continues in its state of rest or of uniform motion in a straight line, unless it is compelled to change that state by forces impressed upon it.
2. The change of motion of an object is proportional to the force impressed and is made in the direction of the straight line in which the force is impressed.
3. To every action there is always opposed an equal reaction; or, the mutual actions of two bodies upon each other are always equal and directed to contrary parts.

Newton's first law is a restatement of the principle of inertia, proposed earlier by Galileo and perfected by Descartes.

The second law is the most important of the three; it may be understood very nearly to summarize all of classical mechanics. Newton used the word "motion" to mean what is today called momentum—that is, the product of mass and velocity, or p = mv, where p is the momentum, m the mass, and v the velocity of a body. The second law may then be written in the form of the equation F = dp/dt, where F is the force, the time derivative expresses Newton's "change of motion," and the vector form of the equation assures that the change is in the same direction as the force, as the second law requires.

For a body whose mass does not change,

$$\frac{dp}{dt} = m\frac{dv}{dt} = ma$$

where a is the acceleration. Thus, Newton's second law may be put in the following form:

$$F = ma$$

It is probably fair to say that equation $F = ma$ is the most famous equation in all of physics.

Newton's third law assures that when two bodies interact, regardless of the nature of the interaction, they do not produce a net force acting on the two-body system as a whole. Instead, there is an action and reaction pair of equal and opposite forces, each acting on a different body (action and reaction forces never act on the same body). The third law applies whether the bodies in question are at rest, in uniform motion, or in accelerated motion.

If a body has a net force acting on it, it undergoes accelerated motion in accordance with the second law. If there is no net force acting on a body, either because there are no forces at all or because all forces are precisely balanced by contrary forces, the body does not accelerate and may be said to be in equilibrium. Conversely, a body that is observed not to be accelerated may be deduced to have no net force acting on it.

Consider, for example, a massive object resting on a table. The object is known to be acted on by the gravitational force of the Earth; if the table were removed, the object would fall. It follows therefore from the fact that the object does not fall that the table exerts an upward force on the object, equal and opposite to the downward force of gravity. This upward force is not a mere physicist's bookkeeping device but rather a real physical force. The table's surface is slightly deformed by the weight of the object, causing the surface to exert a force analogous to that exerted by a coiled spring.

It is useful to recall the following distinction: the massive object exerts a downward force on the table that is equal and opposite to the upward force exerted by the table (owing to its deformation) on the object. These two forces are an action and reaction pair operating on different bodies (one on the table, the other on the object) as required by Newton's third law. On the other hand, the upward force exerted on the object by the table is balanced by a downward force exerted on the object by the Earth's gravity. These two equal and opposite forces, acting on the same body, are not related to or by Newton's third law, but they do produce the equilibrium immobile state of the body.

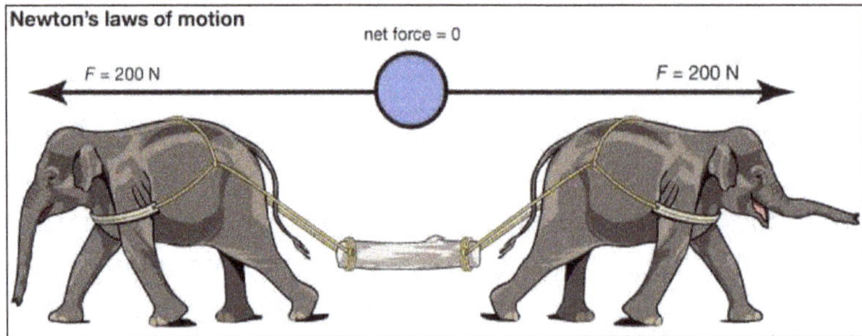

When an object is acted on by equal and opposite forces, it is at rest.

Motion of a Particle in One Dimension

Uniform Motion

According to Newton's first law (also known as the principle of inertia), a body with no net force acting on it will either remain at rest or continue to move with uniform speed in a straight line, according to its initial condition of motion. In fact, in classical Newtonian mechanics, there is no important distinction between rest and uniform motion in a straight line; they may be regarded as the same state of motion seen by different observers, one moving at the same velocity as the particle, the other moving at constant velocity with respect to the particle.

Although the principle of inertia is the starting point and the fundamental assumption of classical mechanics, it is less than intuitively obvious to the untrained eye. In Aristotelian mechanics, and in ordinary experience, objects that are not being pushed tend to come to rest. The law of inertia was deduced by Galileo from his experiments with balls rolling down inclined planes.

For Galileo, the principle of inertia was fundamental to his central scientific task: he had to explain how it is possible that if Earth is really spinning on its axis and orbiting the Sun we do not sense that motion. The principle of inertia helps to provide the answer: Since we are in motion together with Earth, and our natural tendency is to retain that motion, Earth appears to us to be at rest. Thus, the principle of inertia, far from being a statement of the obvious, was once a central issue of scientific contention. By the time Newton had sorted out all the details, it was possible to account accurately for the small deviations from this picture caused by the fact that the motion of Earth's surface is not uniform motion in a straight line. In the Newtonian formulation, the common observation that bodies that are not pushed tend to come to rest is attributed to the fact that they have unbalanced forces acting on them, such as friction and air resistance.

As has already been stated, a body in motion may be said to have momentum equal to the product of its mass and its velocity. It also has a kind of energy that is due entirely to its motion, called kinetic energy. The kinetic energy of a body of mass m in motion with velocity v is given by:

$$K = \frac{1}{2}mv^2$$

Falling Bodies and Uniformly Accelerated Motion

During the 14th century, the French scholar Nicole Oresme studied the mathematical properties of uniformly accelerated motion. He had little interest in whether that kind of motion could be observed in the realm of actual human existence, but he did discover that, if a particle is uniformly accelerated, its speed increases in direct proportion to time, and the distance it traverses is proportional to the square of the time spent accelerating. Two centuries later, Galileo repeated these same mathematical discoveries (perhaps independently) and, just as important, determined that this kind of motion is actually executed by balls rolling down inclined planes. As the incline of the plane increases, the acceleration increases, but the motion continues to be uniformly accelerated. From this observation, Galileo deduced that a body falling freely in the vertical direction would also have uniform acceleration. Even more remarkably, he demonstrated that, in the absence of air resistance, all bodies would fall with the same constant acceleration regardless of their mass. If the constant acceleration of any body dropped near the surface of Earth is expressed as g, the behaviour of a body dropped from rest at height z_0 and time $t = 0$ may be summarized by the following equations:

$$z = z_0 - \frac{1}{2}gt^2$$

$$v = gt$$

$$a = g$$

where z is the height of the body above the surface, v is its speed, and a is its acceleration. These equations of motion hold true until the body actually strikes the surface. The value of g is approximately 9.8 metres per second squared (m/s^2).

A body of mass m at a height z_0 above the surface may be said to possess a kind of energy purely by virtue of its position. This kind of energy (energy of position) is called potential energy. The gravitational potential energy is given by:

$$U = mgz_0.$$

Technically, it is more correct to say that this potential energy is a property of the Earth-body system rather than a property of the body itself, but this pedantic distinction can be ignored.

As the body falls to height z less than z_0, its potential energy U converts to kinetic energy K = $^1/_2mv^2$. Thus, the speed v of the body at any height z is given by solving the equation:

$$\frac{1}{2}mv^2 + mgz = mgz_0.$$

Equation $\frac{1}{2}mv^2 + mgz = mgz_0$ is an expression of the law of conservation of energy. It says that the sum of kinetic energy, $^1/_2mv^2$, and potential energy, mgz, at any point during the fall, is equal to the total initial energy, mgz_0, before the fall began. Exactly the same dependence of speed on height could be deduced from the kinematic equations $z = z_0 - \frac{1}{2}gt^2$, $v = gt$, and $a = g$ above.

In order to reach the initial height z_0, the body had to be given its initial potential energy by some external agency, such as a person lifting it. The process by which a body or a system obtains mechanical energy from outside of itself is called work. The increase of the energy of the body is equal to the work done on it. Work is equal to force times distance.

The force exerted by Earth's gravity on a body of mass m may be deduced from the observation that the body, if released, will fall with acceleration g. Since force is equal to mass times acceleration, the force of gravity is given by F = mg. To lift the body to height z_0, an equal and opposite (i.e., upward) force must be exerted through a distance z_0. Thus, the work done is:

$$W = Fz_0 = mgz_0.$$

which is equal to the potential energy that results.

If work is done by applying a force to a body that is not being acted upon by an opposing force, the body is accelerated. In this case, the work endows the body with kinetic energy rather than potential energy. The energy that the body gains is equal to the work done on it in either case. It should be noted that work, potential energy, and kinetic energy, all being aspects of the same quantity, must all have the dimensions ml^2/t^2.

Simple Harmonic Oscillations

Consider a mass m held in an equilibrium position by springs, as shown in figure. The mass may be perturbed by displacing it to the right or left. If x is the displacement of the mass from equilibrium, the springs exert a force F proportional to x, such that:

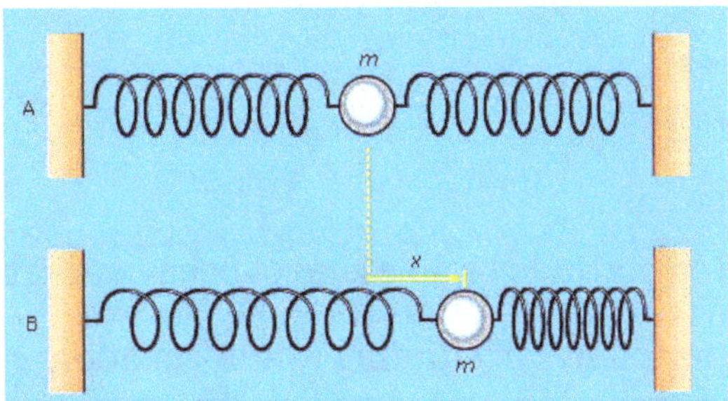

(a) A mass m held in equilibrium by springs. (b) A mass m displaced a distance x.

$$F = -kx,$$

where k is a constant that depends on the stiffness of the springs. Equation $F = -kx$, is called Hooke's law, and the force is called the spring force. If x is positive (displacement to the right), the resulting force is negative (to the left), and vice versa. In other words, the spring force always acts so as to restore mass back toward its equilibrium position. Moreover, the force will produce an acceleration along the x direction given by $a = d^2x/dt^2$. Thus, Newton's second law, F = ma, is applied to this case by substituting −kx for F and d^2x/dt^2 for a, giving $-kx = m(d^2x/dt^2)$. Transposing and dividing by m yields the equation:

$$F = -kx,$$

Equation $a = \frac{d^2x}{dt^2} = -\frac{k}{m}x$ gives the derivative—in this case the second derivative—of a quantity x in terms of the quantity itself. Such an equation is called a differential equation, meaning an equation containing derivatives. Much of the ordinary, day-to-day work of theoretical physics consists of solving differential equations. The question is, given equation $a = \frac{d^2x}{dt^2} = -\frac{k}{m}x$, how does x depend on time?

The answer is suggested by experience. If the mass is displaced and released, it will oscillate back and forth about its equilibrium position. That is, x should be an oscillating function of t, such as a sine wave or a cosine wave. For example, x might obey a behaviour such as:

$$x = A\cos\omega t.$$

Equation $x = A\cos\omega t$. describes the behaviour sketched graphically in figure. The mass is initially displaced a distance x = A and released at time t = 0. As time goes on, the mass oscillates from A to −A and back to A again in the time it takes ωt to advance by 2π. This time is called T, the period of oscillation, so that ωT = 2π, or T = 2π/ω. The reciprocal of the period, or the frequency f, in oscillations per second, is given by f = 1/T = ω/2π. The quantity ω is called the angular frequency and is expressed in radians per second.

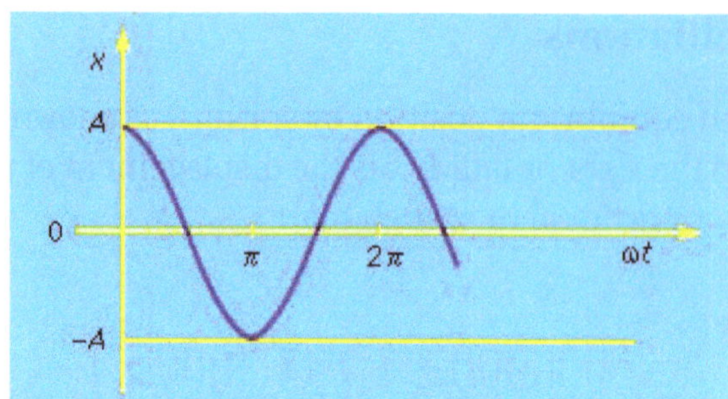

The function x = A cos ωt.

The choice of equation $x = A\cos\omega t$ as a possible kind of behaviour satisfying the differential equation $a=\frac{d^2x}{dt^2}=-\frac{k}{m}x$ can be tested by substituting it into equation $a=\frac{d^2x}{dt^2}=-\frac{k}{m}x$. The first derivative of x with respect to t is:

$$\frac{dx}{dt}=\frac{d}{dt}(A\cos\omega t)$$
$$=-\omega A\sin\omega t$$

Differentiating a second time gives:

$$\frac{d^2x}{dt^2}=\frac{d}{dt}\left(\frac{dx}{dt}\right)$$
$$=\frac{d}{dt}(-\omega A\sin\omega t)$$
$$=-\omega^2 A\cos\omega t$$
$$=-\omega^2 x.$$

Equation $\frac{d^2x}{dt^2}=\frac{d}{dt}\left(\frac{dx}{dt}\right)=\frac{d}{dt}(-\omega A\sin\omega t)=-\omega^2 A\cos\omega t=-\omega^2 x$.is the same as equation $a=\frac{d^2x}{dt^2}=-\frac{k}{m}x.$

If,

$$\omega^2\frac{k}{m}.$$

Thus, subject to this condition, equation $x = A\cos\omega t$ is a correct solution to the differential equation. There are other possible correct guesses (e.g., x = A sin ωt) that differ from this one only in whether the mass is at rest or in motion at the instant t = 0.

The mass, as has been shown, oscillates from A to −A and back again. The speed, given by dx/dt, equation $\frac{dx}{dt}=\frac{d}{dt}(A\cos\omega t)=-\omega A\sin\omega t$, is zero at A and −A, but has its maximum magnitude, equal to ωA, when x is equal to zero. Physically, after the mass is displaced from equilibrium a distance A to the right, the restoring force F pushes the mass back toward its equilibrium position, causing it to accelerate to the left. When it reaches equilibrium, there is no force acting on it at that instant, but it is moving at speed ωA, and its inertia takes it past the equilibrium position. Before it is stopped it reaches position –A, and by this time there is a force acting on it again, pushing it back toward equilibrium.

The whole process, known as simple harmonic motion, repeats itself endlessly with a frequency given by equation $\omega^2 \frac{k}{m}$. Equation $\omega^2 \frac{k}{m}$ means that the stiffer the springs (i.e., the larger k), the higher the frequency (the faster the oscillations). Making the mass greater has exactly the opposite effect, slowing things down.

One of the most important features of harmonic motion is the fact that the frequency of the motion, ω (or f), depends only on the mass and the stiffness of the spring. It does not depend on the amplitude A of the motion. If the amplitude is increased, the mass moves faster, but the time required for a complete round trip remains the same. This fact has profound consequences, governing the nature of music and the principle of accurate timekeeping.

The potential energy of a harmonic oscillator, equal to the work an outside agent must do to push the mass from zero to x, is $U = {}^{1}/_{2}kx^2$ and since the kinetic energy is always ${}^{1}/_{2}mv^2$, when the mass is at any point x in the oscillation,

$$\frac{1}{2}mv^2 + \frac{1}{2}kx^2 = \frac{1}{2}kA^2$$

Equation $\frac{1}{2}mv^2 + \frac{1}{2}kx^2 = \frac{1}{2}kA^2$ plays exactly the role for harmonic oscillators that equation $\frac{1}{2}mv^2 + mgz = mgz_0$. does for falling bodies.

It is quite generally true that harmonic oscillations result from disturbing any body or structure from a state of stable mechanical equilibrium. To understand this point, a brief discussion of stability is useful.

Consider a bowl with a marble resting inside, then consider a second, inverted bowl with a marble balanced on top. In both cases, the net force on the marble is zero. The marbles are thus in mechanical equilibrium. However, a small disturbance in the position of the marble balanced on top of the inverted bowl will cause it to roll away and not return. In such a case, the equilibrium is said to be unstable. Conversely, if the marble inside the first bowl is disturbed, gravity acts to push it back toward the bottom of the bowl. The marble inside the bowl is an example of a body in stable equilibrium. If it is disturbed slightly, it executes harmonic oscillations around the bottom of the bowl rather than rolling away.

This argument may be generalized by a simple mathematical argument. Consider a body or structure in mechanical equilibrium, which, when disturbed by a small amount x, finds a force acting on it that is a function of x, F(x). For small x, such a function may be written generally as a power series in x; i.e.,

$$F(x) = F(0) + ax + bx^2 + ...,$$

where F(o) is the value of F(x) when x = (o), and a and b are constants, independent of x, determined by the nature of the system. The statement that the body is in mechanical equilibrium means that F(o) = o, so that no force is acting on the body when it is undisturbed. Since x is small, x^2 is much smaller; thus, the term bx^2 and all higher powers may be disregarded. This leaves F(x) = ax. Now, if a is positive, a disturbance produces a force in the same direction as the disturbance. This was the case when the marble was balanced on top of the inverted bowl. It describes unstable equilibrium. For the system to be stable, a must be negative. Thus, if a = −k, where k is some positive constant, equation $F(x) = F(0) + ax + bx^2 + ...,$ becomes F(x) = −kx, which is simply Hooke's law, equation $F = -kx$. As has been described above, any system obeying Hooke's law is a harmonic oscillator.

The generality of this argument accounts for the fact that harmonic oscillators are abundantly observed in common experience. For example, any rigid structure will oscillate at many different harmonic frequencies corresponding to different possible distortions of its equilibrium shape. In addition, music may be produced either by disturbing the equilibrium of a stretched wire or fibre (as in the piano and violin), a stretched membrane (e.g., drums), or a rigid bar (the triangle and the xylophone) or by disturbing the density of an enclosed column of air (as in the trumpet and organ). While a fluid such as air is not rigid, its density is an example of a stable system that obeys Hooke's law and may therefore be set into harmonic oscillations.

All music would be quite different from what it is were it not for the general property of harmonic oscillators that the frequency is independent of the amplitude. Thus, instruments yield the same note (frequency) regardless of how loudly they are played (amplitude), and, equally important, the same note persists as the vibrations die away. This same property of harmonic oscillators is the underlying principle of all accurate timekeeping.

The first precise timekeeping mechanism, whose principles of motion were discovered by Galileo, was the simple pendulum. The accuracy of modern timekeeping has been improved dramatically by the introduction of tiny quartz crystals, whose harmonic oscillations generate electrical signals that may be incorporated into miniaturized circuits in clocks and wristwatches. All harmonic oscillators are natural timekeeping devices because they oscillate at intrinsic natural frequencies independent of amplitude. A given number of complete cycles always corresponds to the same elapsed time. Quartz crystal oscillators make more accurate clocks than pendulums do principally because they oscillate many more times per second.

Damped and Forced Oscillations

The simple harmonic oscillations continue forever, at constant amplitude, oscillating as shown in figure between A and −A. Common experience indicates that real oscillators behave somewhat differently, however. Harmonic oscillations tend to die away as time goes on. This behaviour, called damping of the oscillations, is produced by forces such as friction and viscosity. These forces are known collectively as dissipative forces because they tend to dissipate the potential and kinetic energies of macroscopic bodies into the energy of the chaotic motion of atoms and molecules known as heat.

Friction and viscosity are complicated phenomena whose effects cannot be represented accurately by a general equation. However, for slowly moving bodies, the dissipative forces may be represented by:

$$F_d = -\gamma v,$$

where v is the speed of the body and γ is a constant coefficient, independent of dynamic quantities such as speed or displacement. Equation $F_d = -\gamma v,$ is most easily understood by an argument analogous to that applied to equation $F(x) = F(0) + ax + bx^2 + \dots$ F_d is written as a sum of powers of v, or $F_d(v) = F_d(0) + av + bv^2 + \cdots$. When the body is at rest (v = 0), no dissipative force is expected because, if there were one, it might set the body into motion. Thus, $F_d(0) = 0$. The next term must be negative since dissipative forces always resist the motion. Thus, a = −γ where γ is positive. Since v^2 has the same sign regardless of the direction of the motion, b must equal 0 lest it sometimes contribute a dissipative force in the same direction as the motion. The next term is proportional to v^3, and it and all subsequent terms may be neglected if v is sufficiently small. So, as in equation $F(x) = F(0) + ax + bx^2 + \dots,$ the power series is reduced to a single term, in this case $F_d = -\gamma v$.

To find the effect of a dissipative force on a harmonic oscillator, a new differential equation must be solved. The net force, or mass times acceleration, written as md^2x/dt^2, is set equal to the sum of the Hooke's law force, −kx, and the dissipative force, −γv = −γdx/dt. Dividing by m yields:

$$\frac{d^2x}{dt^2} = -\frac{k}{m}x - \frac{\gamma}{m}\frac{dx}{dt}.$$

The general solution to equation $\frac{d^2x}{dt^2} = -\frac{k}{m}x - \frac{\gamma}{m}\frac{dx}{dt}$ is given in the form $x = Ce^{-\gamma t/2m}\cos(\omega t + \theta_0)$, where C and θ_0 are arbitrary constants determined by the initial conditions. This motion, for the case in which $\theta_0 = 0$, is illustrated in figure. As expected, the harmonic oscillations die out with time. The amplitude of the oscillations is bounded by an exponentially decreasing function of time (the dashed curves). The characteristic decay time (after which the oscillations are smaller by 1/e, where e is the base of the natural logarithms e = 2.718) is equal to 2m/γ. The frequency of the oscillations is given by:

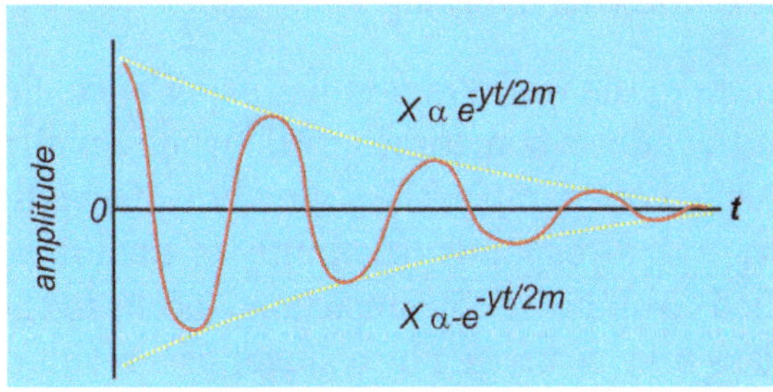

Damped oscillations.

$$\omega^2 = \frac{k}{m} - \frac{\gamma^2}{4m^2}.$$

Importantly, this frequency does not change as the oscillations decay.

Equation $\omega^2 = \frac{k}{m} - \frac{\gamma^2}{4m^2}$ shows that it is possible, by proper choice of γ, to turn a harmonic oscillator into a system that does not oscillate at all—that is, a system whose natural frequency is ω = 0. Such a system is said to be critically damped. For example, the springs that suspend the body of an automobile cause it to be a natural harmonic oscillator. The shock absorbers of the auto are devices that seek to add just enough dissipative force to make the assembly critically damped. In this way, the passengers need not go through numerous oscillations after each bump in the road.

A simple disturbance can set a harmonic oscillator into motion. Repeated disturbances can increase the amplitude of the oscillations if they are applied in synchrony with the natural frequency. Even a very small disturbance, repeated periodically at just the right frequency, can cause a very large amplitude motion to build up. This phenomenon is known as resonance.

Periodically forced oscillations may be represented mathematically by adding a term of the form a_0 sin ωt to the right-hand side of equation $\frac{d^2x}{dt^2} = -\frac{k}{m}x - \frac{\gamma}{m}\frac{dx}{dt}$. This term describes a force applied at frequency ω, with amplitude ma_0. The result of applying such a force is to create a kind of motion that does not need to decay with time, since the energy lost to dissipative processes is replaced, over the course of each cycle, by the driving force. The amplitude of the motion depends on how close the driving frequency ω is to the natural frequency ω_0 of the oscillator. Interestingly, even though dissipation is present, ω_0 is not given by equation $\omega^2 = \frac{k}{m} - \frac{\gamma^2}{4m^2}$ but rather by equation $\omega^2 \frac{k}{m}$: ω^2_0 = k/m. In a graph of the amplitude of the steady state motion (i.e., long after the driving force has begun to be applied), the maximum amplitude occurs as expected at $\omega = \omega_0$. The height and width of the resonance curve are governed by the damping coefficient γ. If there were no damping, the maximum amplitude would be infinite. Because small disturbances at every possible frequency are always present in the natural world, every rigid structure would shake itself to pieces if not for the presence of internal damping.

Resonances are not uncommon in the world of familiar experience. For example, cars often rattle at certain engine speeds, and windows sometimes rattle when an airplane flies by. Resonance is particularly important in music. For example, the sound box of a violin does its job well if it has a natural frequency of oscillation that responds resonantly to each musical note. Very strong resonances to certain notes—called "wolf notes" by musicians—occur in cheap violins and are much to be avoided. Sometimes, a glass may be broken by a singer as a result of its resonant response to a particular musical note.

Motion of a Particle in Two or More Dimensions

Projectile Motion

Galileo was quoted above pointing out with some detectable pride that none before him had realized that the curved path followed by a missile or projectile is a parabola. He had arrived at his conclusion by realizing that a body undergoing ballistic motion executes, quite independently, the motion of a freely falling body in the vertical direction and inertial motion in the horizontal direction. These considerations, and terms such as ballistic and projectile, apply to a body that, once launched, is acted upon by no force other than Earth's gravity.

Projectile motion may be thought of as an example of motion in space—that is to say, of three-dimensional motion rather than motion along a line, or one-dimensional motion. In a suitably defined system of Cartesian coordinates, the position of the projectile at any instant may be specified by giving the values of its three coordinates, x(t), y(t), and z(t). By generally accepted convention, z(t) is used to describe the vertical direction. To a very good approximation, the motion is confined to a single vertical plane, so that for any single projectile it is possible to choose a coordinate system such that the motion is two-dimensional [say, x(t) and z(t)] rather than three-dimensional [x(t), y(t), and z(t)]. It is assumed throughout this topic that the range of the motion is sufficiently limited that the curvature of Earth's surface may be ignored.

Consider a body whose vertical motion obeys equation $z = z_0 - \frac{1}{2}gt^2$, Galileo's law of falling bodies, which states $z = z_0 - {}^1/_2 gt^2$, while, at the same time, moving horizontally at a constant speed vx in accordance with Galileo's law of inertia. The body's horizontal motion is thus described by $x(t) = v_x t$, which may be written in the form $t = x/v_x$. Using this result to eliminate t from equation $z = z_0 - \frac{1}{2}gt^2$ gives $z = z_0 - {}^1/_2 g(1/v_x)^2 x^2$. This latter is the equation of the trajectory of a projectile in the z–x plane, fired horizontally from an initial height z_0. It has the general form:

$$z = a + bx^2,$$

where a and b are constants. Equation $z = a + bx^2$ may be recognized to describe a parabola, just as Galileo claimed. The parabolic shape of the trajectory is preserved even if the motion has an initial component of velocity in the vertical direction.

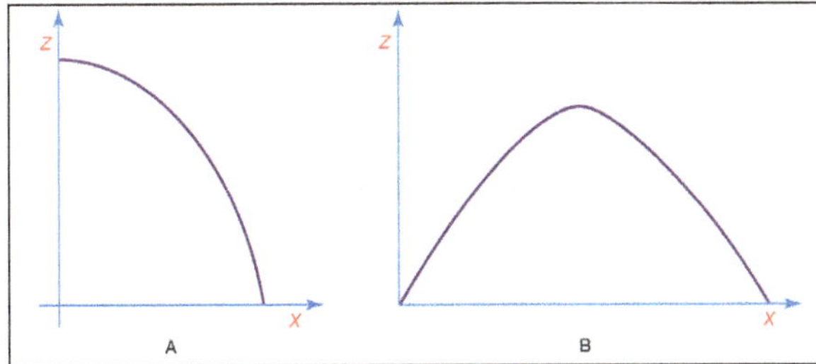

(a) The parabolic path of a projectile. (b) The parabolic path of a projectile with an initial upward component of velocity.

Energy is conserved in projectile motion. The potential energy U(z) of the projectile is given by

$U(z) = mgz$. The kinetic energy K is given by $K = {}^1/_2mv^2$, where v^2 is equal to the sum of the squares of the vertical and horizontal components of velocity, or $v^2 = v^2_x + v^2_z$.

In all of this discussion, the effects of air resistance (to say nothing of wind and other more complicated phenomena) have been neglected. These effects are seldom actually negligible. They are most nearly so for bodies that are heavy and slow-moving. All of this discussion, therefore, is of great value for understanding the underlying principles of projectile motion but of little utility for predicting the actual trajectory of, say, a cannonball once fired or even a well-hit baseball.

Motion of a Pendulum

According to legend, Galileo discovered the principle of the pendulum while attending mass at the Duomo (cathedral) located in the Piazza del Duomo of Pisa, Italy. A lamp hung from the ceiling by a cable and, having just been lit, was swaying back and forth. Galileo realized that each complete cycle of the lamp took the same amount of time, compared to his own pulse, even though the amplitude of each swing was smaller than the last. As has already been shown, this property is common to all harmonic oscillators, and, indeed, Galileo's discovery led directly to the invention of the first accurate mechanical clocks. Galileo was also able to show that the period of oscillation of a simple pendulum is proportional to the square root of its length and does not depend on its mass.

A simple pendulum is sketched in figure. A bob of mass M is suspended by a massless cable or bar of length L from a point about which it pivots freely. The angle between the cable and the vertical is called θ. The force of gravity acting on the mass M, always equal to −Mg in the vertical direction, is a vector that may be resolved into two components, one that acts ineffectually along the cable and another, perpendicular to the cable, that tends to restore the bob to its equilibrium position directly below the point of suspension. This latter component is given by:

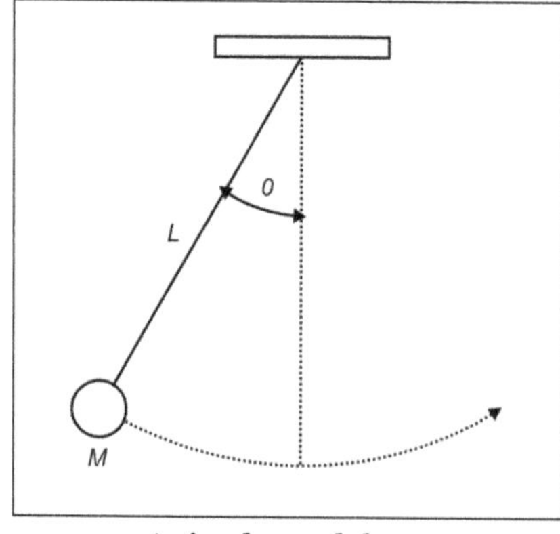

A simple pendulum.

$$F = -Mg\sin\theta.$$

The bob is constrained by the cable to swing through an arc that is actually a segment of a circle of radius L. If the cable is displaced through an angle θ, the bob moves a distance Lθ along its arc (θ must be expressed in radians for this form to be correct). Thus, Newton's second law may be written,

$$F = -Ma = M\frac{d^2(L\theta)}{dt^2}.$$

Equating equation $F = -Mg\sin\theta$ to equation $F = -Ma = M\frac{d^2(L\theta)}{dt^2}$, one sees immediately that the mass M will drop out of the resulting equation. The simple pendulum is an example of a falling body, and its dynamics do not depend on its mass for exactly the same reason that the acceleration of a falling body does not depend on its mass: both the force of gravity and the inertia of the body are proportional to the same mass, and the effects cancel one another. The equation that results (after extracting the constant L from the derivative and dividing both sides by L) is

$$\frac{d^2\theta}{dt^2} = -\frac{g}{L}\sin\theta.$$

If the angle θ is sufficiently small, equation $\frac{d^2\theta}{dt^2} = -\frac{g}{L}\sin\theta$ may be rewritten in a form that is both more familiar and more amenable to solution. Figure shows a segment of a circle of radius L. A radius vector at angle θ, as shown, locates a point on the circle displaced a distance Lθ along the arc. It is clear from the geometry that L sin θ and Lθ are very nearly equal for small θ. It follows then that sin θ and θ are also very nearly equal for small θ. Thus, if the analysis is restricted to small angles, then sin θ may be replaced by θ in equation $\frac{d^2\theta}{dt^2} = -\frac{g}{L}\sin\theta$ to obtain:

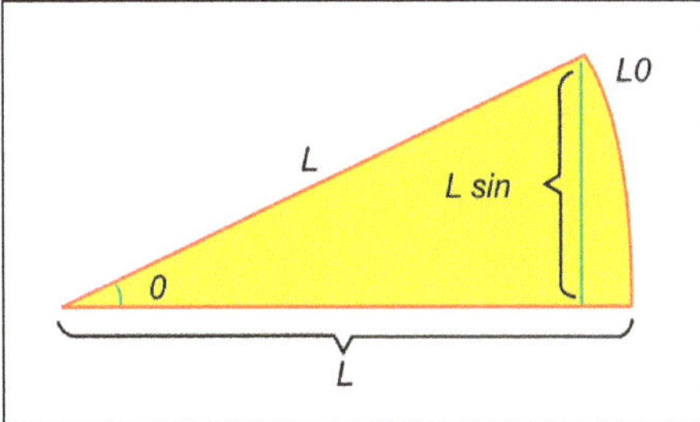

A segment of a circle of radius L.

$$\frac{d^2\theta}{dt^2} = -\frac{g}{L}\theta.$$

Equation $\frac{d^2\theta}{dt^2} = -\frac{g}{L}\theta$ should be compared with equation $a = \frac{d^2x}{dt^2} = -\frac{k}{m}x$: $d^2x/dt^2 = -(k/m)x$. In the first case, the dynamic variable (meaning the quantity that changes with time) is θ, in the second case it is x. In both cases, the second derivative of the dynamic variable with respect to time is equal to the variable itself multiplied by a negative constant. The equations are therefore mathematically identical and have the same solution—i.e., equation $x = A\cos\omega t$, or θ = A cos ωt.

In the case of the pendulum, the frequency of the oscillations is given by the constant in equation $\frac{d^2\theta}{dt^2} = -\frac{g}{L}\theta$, or $\omega^2 = g/L$. The period of oscillation, T = 2π/ω, is therefore,

$$T = 2\pi\sqrt{\frac{L}{g}}.$$

Just as Galileo concluded, the period is independent of the mass and proportional to the square root of the length.

As with most problems in physics, this discussion of the pendulum has involved a number of simplifications and approximations. Most obviously, sin θ was replaced by θ to obtain $\frac{d^2\theta}{dt^2} = -\frac{g}{L}\theta.$ This approximation is surprisingly accurate. For example, at a not-very-small angle of 17.2°, corresponding to 0.300 radian, sin θ is equal to 0.296, an error of less than 2 percent. For smaller angles, of course, the error is appreciably smaller.

The problem was also treated as if all the mass of the pendulum were concentrated at a point at the end of the cable. This approximation assumes that the mass of the bob at the end of the cable is much larger than that of the cable and that the physical size of the bob is small compared with the length of the cable. When these approximations are not sufficient, one must take into account the way in which mass is distributed in the cable and bob. This is called the physical pendulum, as opposed to the idealized model of the simple pendulum. Significantly, the period of a physical pendulum does not depend on its total mass either.

The effects of friction, air resistance, and the like have also been ignored. These dissipative forces have the same effects on the pendulum as they do on any other kind of harmonic oscillator, as. They cause the amplitude of a freely swinging pendulum to grow smaller on successive swings. Conversely, in order to keep a pendulum clock going, a mechanism is needed to restore the energy lost to dissipative forces.

Circular Motion

Consider a particle moving along the perimeter of a circle at a uniform rate, such that it makes one complete revolution every hour. To describe the motion mathematically, a vector is constructed from the centre of the circle to the particle. The vector then makes one complete revolution every hour. In other words, the vector behaves exactly like the large hand on a wristwatch, an arrow of fixed length that makes one complete revolution every hour. The motion of the point of the vector is an example of uniform circular motion, and the period T of the motion is equal to one hour (T = 1 h). The arrow sweeps out an angle of 2π radians (one complete circle) per hour. This rate is called the angular frequency and is written $\omega = 2\pi\ h^{-1}$. Quite generally, for uniform circular motion at any rate,

$$T = \frac{2\pi}{\omega}.$$

These definitions and relations are the same as they are for harmonic motion.

Consider a coordinate system, as shown in figure, with the circle centred at the origin. At any instant of time, the position of the particle may be specified by giving the radius r of the circle and the angle θ between the position vector and the x-axis. Although r is constant, θ increases uniformly with time t, such that θ = ωt, or dθ/dt = ω, where ω is the angular frequency in equation $T = \frac{2\pi}{\omega}$. Contrary to the case of the wristwatch, however, ω is positive by convention when the rotation is in the counterclockwise sense. The vector r has x and y components given by:

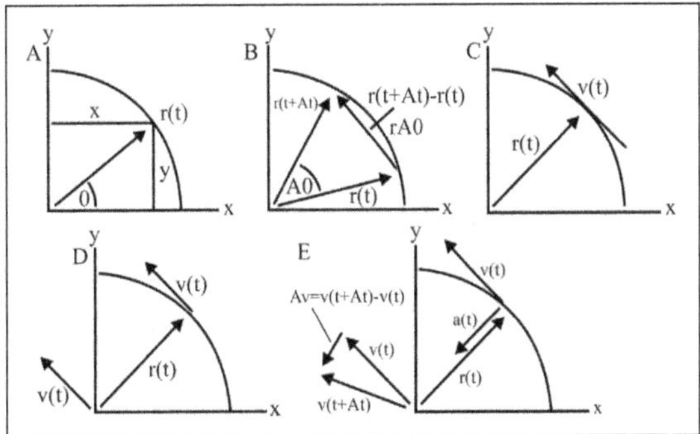

(A) A coordinate system to describe uniform circular motion. (B) The distance traveled in time Δt by a particle undergoing uniform circular motion. (C) The instantaneous velocity of the particle. (D) The velocity vector v undergoes uniform circular motion at the same angular frequency as the particle. (E) The acceleration vector of the particle.

$$x = r\cos\theta = r\cos\omega t,$$

$$y = r\sin\theta = r\sin\omega t.$$

One meaning of equations $x = r\cos\theta = r\cos\omega t$ and $y = r\sin\theta = r\sin\omega t$ is that, when a particle undergoes uniform circular motion, its x and y components each undergo simple harmonic motion. They are, however, not in phase with one another: at the instant when x has its maximum amplitude (say, at θ = 0), y has zero amplitude, and vice versa.

In a short time, Δt, the particle moves rΔθ along the circumference of the circle, as shown in figure. The average speed of the particle is thus given by:

$$\bar{v} = r\frac{\Delta\theta}{\Delta t}.$$

The average velocity of the particle is a vector given by:

$$\bar{v} = \frac{r(t+\Delta t) - r(t)}{\Delta t}.$$

This operation of vector subtraction is indicated in figure. It yields a vector that is nearly perpendicular to r(t) and r(t + Δt). Indeed, the instantaneous velocity, found by allowing Δt to shrink to zero, is a vector v that is perpendicular to r at every instant and whose magnitude is:

$$|v| = r\frac{d\theta}{dt} = r\omega.$$

The relationship between r and v is shown in figure. It means that the particle's instantaneous velocity is always tangent to the circle.

Notice that, just as the position vector r may be described in terms of the components x and y given by equations $x = r\cos\theta = r\cos\omega t$ and $y = r\sin\theta = r\sin\omega t$, the velocity vector v may be described in terms of its projections on the x and y axes, given by:

$$v_x = \frac{dx}{dt} = -r\omega \sin \omega t,$$

$$v_y = \frac{dy}{dt} = r\omega \cos \omega t$$

Imagine a new coordinate system, in which a vector of length ωr extends from the origin and points at all times in the same direction as v. This construction is shown in figure. Each time the particle sweeps out a complete circle, this vector also sweeps out a complete circle. In fact, its point is executing uniform circular motion at the same angular frequency as the particle itself. Because vectors have magnitude and direction, but not position in space, the vector that has been constructed is the velocity v. The velocity of the particle is itself undergoing uniform circular motion at angular frequency ω.

Although the speed of the particle is constant, the particle is nevertheless accelerated, because its velocity is constantly changing direction. The acceleration a is given by:

$$a = \frac{dv}{dt}.$$

Since v is a vector of length rω undergoing uniform circular motion, equations $\bar{v} = r\frac{\Delta\theta}{\Delta t}$ and $\bar{v} = \frac{r(t+\Delta t) - r(t)}{\Delta t}$ may be repeated, as illustrated in figure, giving:

$$\bar{a} = r\omega\frac{\Delta\theta}{\Delta t}$$

$$\bar{a} = \frac{v(t+\Delta t) - v(t)}{\Delta t}.$$

Thus, one may conclude that the instantaneous acceleration is always perpendicular to v and its magnitude is:

$$|a| = r\omega\frac{}{dt} = r\omega$$

Since v is perpendicular to r, and a is perpendicular to v, the vector a is rotated 180° with respect to r. In other words, the acceleration is parallel to r but in the opposite direction. The same conclusion may be reached by realizing that a has x and y components given by:

$$a_x = \frac{dv_x}{dt} = -r\omega^2 \cos \omega t,$$

$$a_y = \frac{dv_y}{dt} = -r\omega^2 \sin \omega t,$$

similar to equations $v_x = \frac{dx}{dt} = -r\omega \sin \omega t,$ and $v_y = \frac{dy}{dt} = r\omega \cos \omega t$. When equations $a_x = \frac{dv_x}{dt} = -r\omega^2 \cos \omega t,$ and $a_y = \frac{dv_y}{dt} = -r\omega^2 \sin \omega t,$ are compared with equations $x = r\cos\theta = r\cos\omega t,$ and $y = r\sin\theta = r\sin\omega t.$ for x and y, it is clear that the components of a are just those of r multiplied by $-\omega^2$, so that $a = -\omega^2 r$. This acceleration is called the centripetal acceleration, meaning that it is inward, pointing along the radius vector toward the centre of the circle. It is sometimes useful to express the centripetal acceleration in terms of the speed v. Using $v = \omega r$, one can write:

$$a = -\frac{v^2}{r}.$$

Circular Orbits

The detailed behaviour of real orbits is the concern of celestial mechanics. In fact, Earth's orbit about the Sun is not quite exactly uniformly circular, but it is a close enough approximation for the purposes of this discussion.

A body in uniform circular motion undergoes at all times a centripetal acceleration given by equation $a = -\frac{v^2}{r}$. According to Newton's second law, a force is required to produce this acceleration. In the case of an orbiting planet, the force is gravity. The gravitational attraction of the Sun is an inward (centripetal) force acting on Earth. This force produces the centripetal acceleration of the orbital motion.

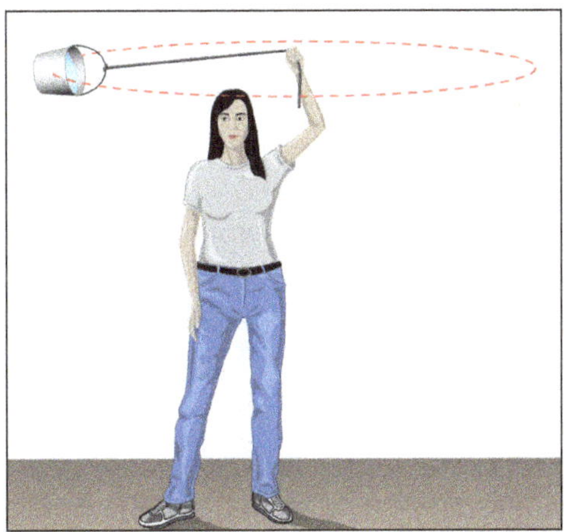

The bucket experiences a centripetal force directed along the string toward the centre of the circle.

Before these ideas are expressed quantitatively, an understanding of why a force is needed to maintain a body in an orbit of constant speed is useful. The reason is that, at each instant, the velocity of the planet is tangent to the orbit. In the absence of gravity, the planet would obey the law of inertia (Newton's first law) and fly off in a straight line in the direction of the velocity at constant speed. The force of gravity serves to overcome the inertial tendency of the planet, thereby keeping it in orbit.

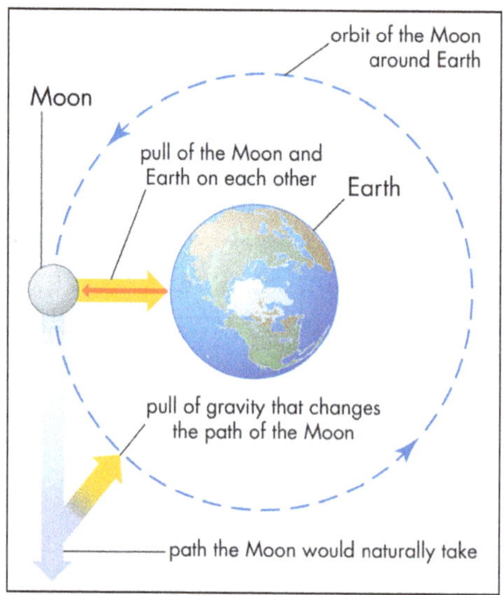

Effects of gravity on Earth and the Moon.

The gravitational force between two bodies such as the Sun and Earth is given by:

$$F = -G\frac{M_s M_E}{r^2},$$

where M_S and M_E are the masses of the Sun and Earth, respectively, r is the distance between their centres, and G is a universal constant equal to 6.674×10^{-11} Nm^2/kg^2 (Newton metres squared per kilogram squared). The force acts along the direction connecting the two bodies (i.e., along the radius vector of the uniform circular motion), and the minus sign signifies that the force is attractive, acting to pull Earth toward the Sun.

To an observer on the surface of Earth, the planet appears to be at rest at (approximately) a constant distance from the Sun. It would appear to the observer, therefore, that any force (such as the Sun's gravity) acting on Earth must be balanced by an equal and opposite force that keeps Earth in equilibrium. In other words, if gravity is trying to pull Earth into the Sun, some opposing force must be present to prevent that from happening. In reality, no such force exists. Earth is in freely accelerated motion caused by an unbalanced force. The apparent force, known in mechanics as a pseudoforce, is due to the fact that the observer is actually in accelerated motion. In the case of orbital motion, the outward pseudoforce that balances gravity is called the centrifugal force.

For a uniform circular orbit, gravity produces an inward acceleration given by equation $a = -\frac{v^2}{r}$, $a = -v^2/r$. The pseudoforce f needed to balance this acceleration is just equal to the mass of Earth times an equal and opposite acceleration, or $f = M_E v^2/r$. The earthbound observer then believes that there is no net force acting on the planet—i.e., that $F + f = 0$, where F is the force of gravity given by equation $F = -G\frac{M_s M_E}{r^2}$. Combining these equations yields a relation between the speed v of a planet and its distance r from the Sun:

$$v^2 = G\frac{M_s}{r},$$

It should be noted that the speed does not depend on the mass of the planet. This occurs for exactly the same reason that all bodies fall toward Earth with the same acceleration and that the period of a pendulum is independent of its mass. An orbiting planet is in fact a freely falling body.

Equation $v^2 = G\frac{M_s}{r}$, is a special case (for circular orbits) of Kepler's third law, Using the fact that v = $2\pi r/T$, where $2\pi r$ is the circumference of the orbit and T is the time to make a complete orbit (i.e., T is one year in the life of the planet), it is easy to show that $T^2 = (4\pi^2/GM_S)r^3$. This relation also may be applied to satellites in circular orbit around Earth (in which case, M_E must be substituted for M_S) or in orbit around any other central body.

Angular Momentum and Torque

A particle of mass m and velocity v has linear momentum p = mv. The particle may also have angular momentum L with respect to a given point in space. If r is the vector from the point to the particle, then,

$$L = r \times P.$$

Notice that angular momentum is always a vector perpendicular to the plane defined by the vectors r and p (or v). For example, if the particle (or a planet) is in a circular orbit, its angular momentum with respect to the centre of the circle is perpendicular to the plane of the orbit and in the direction given by the vector cross product right-hand rule, as shown in figure. Moreover, since in the case of a circular orbit, r is perpendicular to p (or v), the magnitude of L is simply:

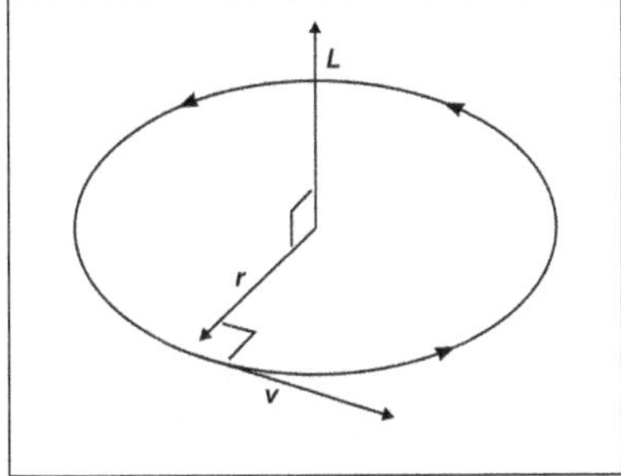

The angular momentum L of a particle traveling in a circular orbit.

$$L = rp = mvr.$$

The significance of angular momentum arises from its derivative with respect to time,

$$\frac{dL}{dt} = \frac{d}{dt}(r \times p) = m\frac{d}{dt}(r \times v).$$

where p has been replaced by mv and the constant m has been factored out. Using the product rule of differential calculus,

$$\frac{d}{dt}(r \times v) = \frac{dr}{dt} \times v + r \times \frac{dv}{dt}.$$

In the first term on the right-hand side of equation $\frac{d}{dt}(r \times v) = \frac{dr}{dt} \times v + r \times \frac{dv}{dt}$, dr/dt is simply the velocity v, leaving v × v. Since the cross product of any vector with itself is always zero, that term drops out, leaving:

$$\frac{d}{dt}(r \times v) = r \times \frac{dv}{dt}.$$

Here, dv/dt is the acceleration a of the particle. Thus, if equation $\frac{d}{dt}(r \times v) = r \times \frac{dv}{dt}$ is multiplied by m, the left-hand side becomes dL/dt, as in equation $\frac{dL}{dt} = \frac{d}{dt}(r \times p) = m\frac{d}{dt}(r \times v)$, and the right-hand side may be written r × ma. Since, according to Newton's second law, ma is equal to F, the net force acting on the particle, the result is:

$$\frac{dL}{dt} = r \times F.$$

Equation $\frac{dL}{dt} = r \times F$ means that any change in the angular momentum of a particle must be produced by a force that is not acting along the same direction as r. One particularly important application is the solar system. Each planet is held in its orbit by its gravitational attraction to the Sun, a force that acts along the vector from the Sun to the planet. Thus, the force of gravity cannot change the angular momentum of any planet with respect to the Sun. Therefore, each planet has constant angular momentum with respect to the Sun. This conclusion is correct even though the real orbits of the planets are not circles but ellipses.

$$\frac{dL}{dt} = r \times F.$$

The quantity r × F is called the torque τ. Torque may be thought of as a kind of twisting force, the kind needed to tighten a bolt or to set a body into rotation. Using this definition, equation $\frac{dL}{dt} = r \times F.$ may be rewritten

Equation $\tau = r \times F = \frac{dL}{dt}$ means that if there is no torque acting on a particle, its angular momentum is constant, or conserved. Suppose, however, that some agent applies a force F_a to the particle resulting in a torque equal to r × F_a. According to Newton's third law, the particle must apply a force $-F_a$ to the agent. Thus, there is a torque equal to $-r \times F_a$ acting on the agent. The torque on the particle causes its angular momentum to change at a rate given by $dL/dt = r \times F_a$. However, the angular momentum L_a of the agent is changing at the rate $dL_a/dt = -r \times F_a$. Therefore, $dL/dt + dL_a/dt = 0$, meaning that the total angular momentum of particle plus agent is constant, or conserved. This principle may be generalized to include all interactions between bodies of any kind, acting by way of forces of any kind. Total angular momentum is always conserved. The law of conservation of angular momentum is one of the most important principles in all of physics.

Motion of a Group of Particles

Centre of Mass

The word particle has been used in this article to signify an object whose entire mass is concentrated at a point in space. In the real world, however, there are no particles of this kind. All real bodies have sizes and shapes. Furthermore, as Newton believed and is now known, all bodies are in fact compounded of smaller bodies called atoms. Therefore, the science of mechanics must deal not only with particles but also with more complex bodies that may be thought of as collections of particles.

To take a specific example, the orbit of a planet around the Sun was discussed earlier as if the planet and the Sun were each concentrated at a point in space. In reality, of course, each is a substantial body. However, because each is nearly spherical in shape, it turns out to be permissible, for the purposes of this problem, to treat each body as if its mass were concentrated at its centre. This is an example of an idea that is often useful in discussing bodies of all kinds: the centre of mass. The centre of mass of a uniform sphere is located at the centre of the sphere. For many purposes (such as the one cited above) the sphere may be treated as if all its mass were concentrated at its centre of mass.

To extend the idea farther, consider Earth and the Sun not as two separate bodies but as a single system of two bodies interacting with one another by means of the force of gravity. In the previous discussion of circular orbits, the Sun was assumed to be at rest at the centre of the orbit, but, according to Newton's third law, it must actually be accelerated by a force due to Earth that is equal and opposite to the force that the Sun exerts on Earth. In other words, considering only the Sun and Earth (ignoring, for example, all the other planets), if M_S and M_E are, respectively, the masses of the Sun and Earth, and if a_S and a_E are their respective accelerations, then combining Newton's second and third laws results in the equation $M_S a_S = -M_E a_E$. Writing each a as dv/dt, this equation is easily manipulated to give:

$$\frac{d}{dt}(M_S v_S + M_E v_E) = 0,$$

$$M_S v_S + M_E v_E = \text{constant}.$$

This remarkable result means that, as Earth orbits the Sun and the Sun moves in response to Earth's gravitational attraction, the entire two-body system has constant linear momentum, moving in a straight line at constant speed. Without any loss of generality, one can imagine observing the system from a frame of reference moving along with that same speed and direction. This is sometimes called the centre-of-mass frame. In this frame, the momentum of the two-body system—i.e., the constant in equation $M_S v_S + M_E v_E = \text{constant}$—is equal to zero. Writing each of the v's as the corresponding dr/dt, equation $M_S v_S + M_E v_E = \text{constant}$. may be expressed in the form

$$M_S v_S + M_E v_E = \text{constant}.$$

$$\frac{d}{dt}(M_S r_S + M_E r_E) = 0,$$

Thus, $M_S r_S$ and $M_E r_E$ are two vectors whose vector sum does not change with time. The sum is defined to be the constant vector MR, where M is the total mass of the system and equals $M_S + M_E$. Thus,

$$MR = M_S r_S + M_E r_E.$$

This procedure defines a constant vector R, from any arbitrarily chosen point in space. The relation between vectors R, r_S, and r_E is shown in figure. The fact that R is constant (although r_S and r_E are not constant) means that, rather than Earth orbiting the Sun, Earth and the Sun are both orbiting an imaginary point fixed in space. This point is known as the centre of mass of the two-body system.

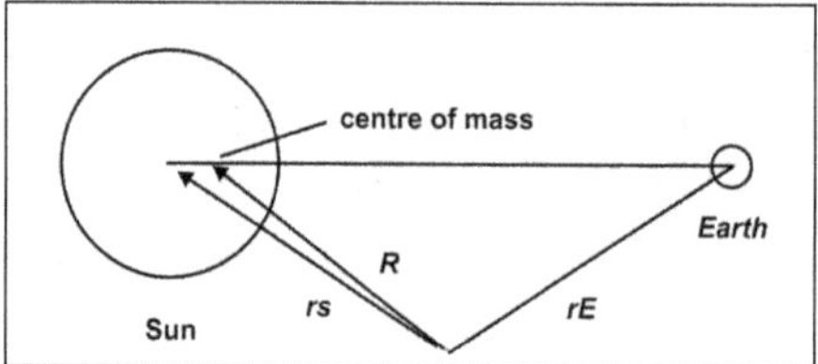

The centre of mass of the two-body Earth-Sun system.

Knowing the masses of the two bodies (MS = 1.99×10^{30} kilograms, ME = 5.98×10^{24} kilograms), it is easy to find the position of the centre of mass. The origin of the coordinate system may be chosen to be located at the centre of mass merely by defining R = 0. Then rS = (ME/MS) rE ≈ 450 kilometres, when rE is rounded to 1.5×10^8 km. A few hundred kilometres is so small compared to rE that, for all practical purposes, no appreciable error occurs when rS is ignored and the Sun is assumed to be stationary at the centre of the orbit.

With this example as a guide, it is now possible to define the centre of mass of any collection of bodies. Assume that there are N bodies altogether, each labeled with numbers ranging from 1 to N, and that the vector from an arbitrary origin to the ith body—where i is some number between 1 and N—is r_i, as shown in figure. Let the mass of the ith body be m_i. Then the total mass of the N-body system is:

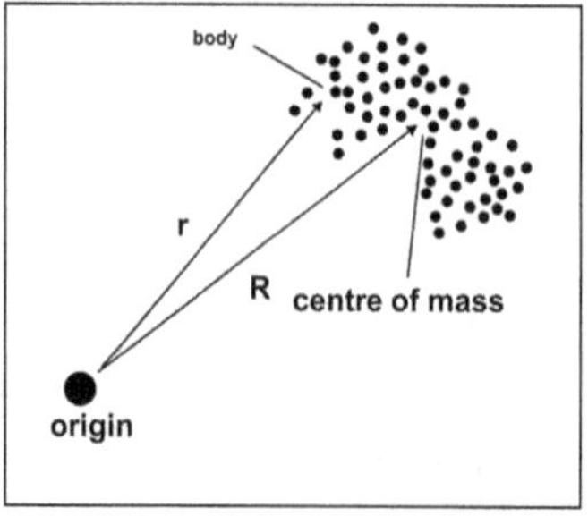

The centre of mass of an N-body system.

$$m = \sum_{i=1}^{N} m_i,$$

and the centre of mass of the system is found at the end of a vector R given by:

$$mR = \sum_{i=1}^{N} m_i r_i,$$

as illustrated in figure. This definition applies regardless of whether the N bodies making up the system are the stars in a galaxy, the atoms in a rigid body, larger and arbitrarily chosen segments of a rigid body, or any other system of masses. According to equation $mR = \sum_{i=1}^{N} m_i r_i,$, the vector to the centre of mass of any system is a kind of weighted average of the vectors to all the components of the system.

As will be demonstrated in the sections that follow, the statics and dynamics of many complicated bodies or systems may often be understood by simply applying Newton's laws as if the system's mass were concentrated at the centre of mass.

Conservation of Momentum

Newton's second law, in its most general form, says that the rate of a change of a particle's momentum p is given by the force acting on the particle; i.e., F = dp/dt. If there is no force acting on the particle, then, since dp/dt = 0, p must be constant, or conserved. This observation is merely a restatement of Newton's first law, the principle of inertia: if there is no force acting on a body, it moves at constant speed in a straight line.

Now suppose that an external agent applies a force F_a to the particle so that p changes according to:

$$\frac{dp}{dt} = F_a.$$

According to Newton's third law, the particle must apply an equal and opposite force $-F_a$ to the external agent. The momentum pa of the external agent therefore changes according to:

$$\frac{dp_a}{dt} = -F_a.$$

Adding together equations $\frac{dp}{dt} = F_a$ and $\frac{dp_a}{dt} = -F_a$ results in the equation,

$$\frac{d}{dt}(p + p_a) = 0.$$

The force applied by the external agent changes the momentum of the particle, but at the same time the momentum of the external agent must also change in such a way that the total momentum of both together is constant, or conserved. This idea may be generalized to give the law of conservation of momentum: in all the interactions between all the bodies in the universe, total momentum is always conserved.

It is useful in this light to examine the behaviour of a complicated system of many parts. The centre of mass of the system may be found using equation $mR = \sum_{i=1}^{N} m_i r_i$. Differentiating with respect to time gives:

$$mv = \sum_{i=1}^{N} m_i v_i$$

where v = dR/dt and $v_i = dr_i/dt$. Note that mivi is the momentum of the ith part of the system, and mv is the momentum that the system would have if all its mass (i.e., m) were concentrated at its centre of mass, the point whose velocity is v. Thus, the momentum associated with the centre of mass is the sum of the momenta of the parts.

Suppose now that there is no external agent applying a force to the entire system. Then the only forces acting on the system are those exerted by the parts on one another. These forces may accelerate the individual parts. Differentiating equation $mv = \sum_{i=1}^{N} m_i v_i$ with respect to time gives:

$$m\frac{dv}{dt} = \sum_{i=1}^{N} m_i \frac{dv_i}{dt} = \sum_{i=1}^{N} F_i,$$

where F_i is the net force, or the sum of the forces, exerted by all the other parts of the body on the ith part. F_i is defined mathematically by the equation:

$$F_i = \sum_{j=1}^{N} F_{ij},$$

where F_{ij} represents the force on body i due to body j (the force on body i due to itself, F_{ii}, is zero). The motion of the centre of mass is then given by the complicated-looking formula:

$$m\frac{dv}{dt} = \sum_{i=1}^{N} \left(\sum_{j=1}^{N} F_{ij} \right).$$

This complicated formula may be greatly simplified, however, by noting that Newton's third law requires that for every force F_{ij} exerted by the jth body on the ith body, there is an equal and opposite force $-F_{ij}$ exerted by the ith body on the jth body. In other words, every term in the double sum has an equal and opposite term. The double summation on the right-hand side of equation $F_i = \sum_{j=1}^{N} F_{ij}$, always adds up to zero. This result is true regardless of the complexity of the system, the nature of the forces acting between the parts, or the motions of the parts. In short, in the absence of external forces acting on the system as a whole, mdv/dt = 0, which means that the momentum of the centre of mass of the system is always conserved. Having determined that momentum is conserved whether or not there is an external force acting, one may conclude that the total momentum of the universe is always conserved.

Collisions

A collision is an encounter between two bodies that alters at least one of their courses. Altering the course of a body requires that a force be applied to it. Thus, each body exerts a force on the other. These forces of interaction may operate at some distance, as do the gravitational and electromagnetic forces, or the bodies may appear to make physical contact. However, even apparent contact between two bodies is only a macroscopic manifestation of microscopic forces that act between atoms some distance apart. There is no fundamental distinction between physical contact and interaction at a distance.

The importance of understanding the mechanics of collisions is obvious to anyone who has ever driven an automobile. In modern physics, however, collisions are important for a different reason. The current understanding of the subatomic particles of which atoms are composed is derived entirely from studying the results of collisions among them. Thus, in modern physics, the description of collisions is a significant part of the understanding of matter. These descriptions are quantum mechanical rather than classical, but they are nevertheless closely based on principles that arise out of classical mechanics.

It is possible in principle to predict the result of a collision using Newton's second law directly. Suppose that two bodies are going to collide and that F, the force of interaction between them, is known to be a function of r, the distance between them. Then, if it is known that, say, one particle has incident momentum p, the problem is solved if the final momentum p + Δp can be determined. Inverting Newton's second law, F = dp/dt, the change in momentum is given by:

$$\Delta p = \int_{-\infty}^{\infty} F dt.$$

This integral is known as the impulse imparted to the particle. In order to perform the integral, it is necessary to know r at all times so that F may be known at all times. More realistically, Δp is the sum of a series of small steps, such that:

$$\delta p = F \delta t,$$

where F depends on the instantaneous distance between the particles. Because p = mv = mdr/dt, the change in r in this step is:

$$\delta r = \frac{p}{m} \delta t.$$

At the next step, there is a new distance, r + δr, giving a new value of the force in equation $\delta p = F \delta t$, and a new momentum, p + δp, in equation $\delta r = \frac{p}{m} \delta t$. This method of analyzing collisions is used in numerical calculations on digital computers.

To predict the result of a collision analytically (rather than numerically) it is often most useful to apply conservation laws. In any collision (as in any other phenomenon), energy, momentum, and angular momentum are always conserved. Judicious application of these laws may be extremely useful because they do not depend in any way on the detailed nature of the interaction (i.e., the force as a function of distance).

This point can be illustrated by the following example. A collision is to take place between two bodies of the same mass m. One of the bodies is initially at rest (its momentum is zero). The other has initial momentum p_0. After the collision, the body previously at rest has momentum p_1, and the body initially in motion has momentum p_2. Since momentum is conserved, the total momentum after the collision, $p_1 + p_2$, must be equal to the total momentum before the collision, p_0; that is:

$$P_0 = P_1 + P_2.$$

Equation $P_0 = P_1 + P_2$ is the equation of a vector triangle, as shown in figure. However, p_1 and p_2 are not determined by this condition; they are only constrained by it.

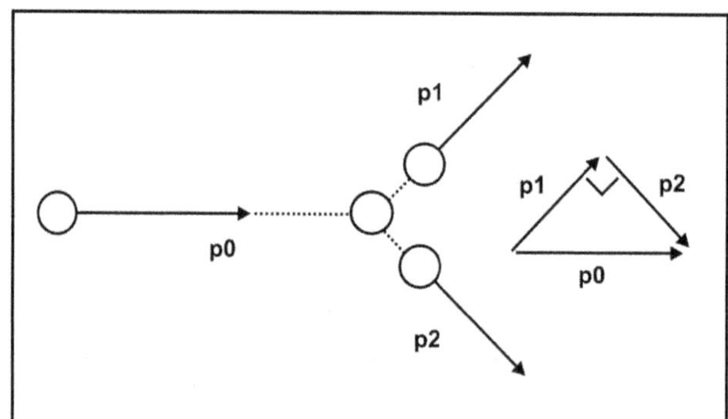

Collision between two particles of equal mass.

Although energy is always conserved, the kinetic energy of the incident body is not always converted entirely into the kinetic energy of the two bodies after the collision. For example, if the bodies are microscopic (say, two identical atoms), the collision may cause one or both to be excited into a state of higher internal energy than it started with. Such an event would leave correspondingly less kinetic energy for the outgoing atoms. In fact, it is precisely by studying the trajectories of outgoing projectiles in collisions like these that physicists are able to determine the possible excited states of microscopic particles.

In a collision between macroscopic objects, some of the kinetic energy is always converted to heat. Heat is the energy of random vibrations of the atoms and molecules that constitute the bodies. However, if the amount of heat is negligible compared to the initial kinetic energy, it may be ignored. Such a collision is said to be elastic.

Suppose the collision described above between two bodies, each of mass m, is between billiard balls, and suppose it is elastic (a reasonably good approximation of real billiard balls). The kinetic energy of the incident ball is then equal to the sum of the kinetic energies of the outgoing balls. According to equation $K = \frac{1}{2}mv^2$, the kinetic energy of a moving object is given by $K = {}^1/_2mv^2$, where v is the speed of the ball. Equation $K = \frac{1}{2}mv^2$ may be written in a particularly useful form by recognizing that since p = mv,

$$K = \frac{1}{2}mv^2 = \frac{p^2}{2m}.$$

Then the conservation of kinetic energy may be written,

$$\frac{p_0^2}{2m} = \frac{p_1^2}{2m} + \frac{p_2^2}{2m},$$

or, canceling the factors 2m,

$$p_0^2 = p_1^2 + p_2^2$$

Comparing this result with equation $P_0 = P_1 + P_2$ shows that the vector triangle is pythagorean; p_1 and p_2 are perpendicular. This result is well known to all experienced pool players. Notice that it was possible to arrive at this result without any knowledge of the forces that act when billiard balls collide.

Relative Motion

A collision between two bodies can always be described in a frame of reference in which the total momentum is zero. This is the centre-of-mass (or centre-of-momentum) frame mentioned earlier. Then, for example, in the collision between two bodies of the same mass, the two bodies always have equal and opposite velocities, as shown in figure. It should be noted that, in this frame of reference, the outgoing momenta are antiparallel and not perpendicular.

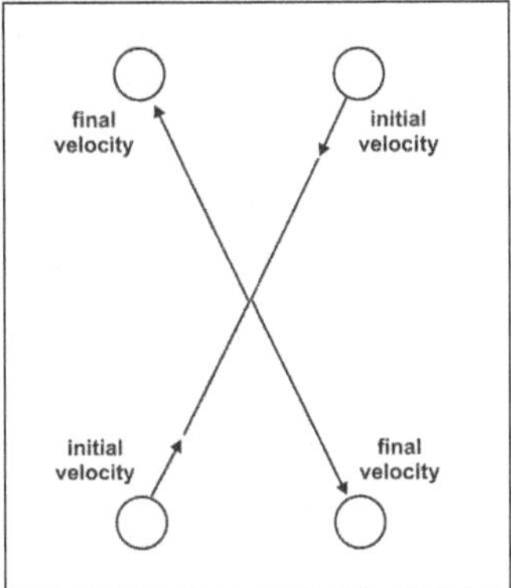

Collision between two particles of equal mass as seen from the centre-of-mass frame of reference.

Any collection of bodies may similarly be described in a frame of reference in which the total momentum is zero. This frame is simply the one in which the centre of mass is at rest. This fact is easily seen by differentiating equation $mR = \sum_{i=1}^{N} m_i r_i$, with respect to time, giving:

$$m\frac{dR}{dt} = \sum_{i=1}^{N} m_i \frac{dr_i}{dt}.$$

The right-hand side is the sum of the momenta of all the bodies. It is equal to zero if the velocity of the centre of mass, dR/dt, is equal to zero.

If Newton's second law is correct in any frame of reference, it will also appear to be correct to an observer moving with any constant velocity with respect to that frame. This principle, called the principle of Galilean relativity, is true because, to the moving observer, the same constant velocity seems to have been added to the velocity of every particle in the system. This change does not affect the accelerations of the particles (since the added velocity is constant, not accelerated) and therefore does not change the apparent force (mass times acceleration) acting on each particle. That is why it is permissible to describe a problem from the centre-of-momentum frame (provided that the centre of mass is not accelerated) or from any other frame moving at constant velocity with respect to it.

If this principle is strictly correct, the fundamental forces of physics should not contain any particular speed. This must be true because the speed of any object will be different to observers in different but equally good frames of reference, but the force should always be the same. It turns out, according to the theory of James Clerk Maxwell, that there is an intrinsic speed in the force laws of electricity and magnetism: the speed of light appears in the forces between electric charges and between magnetic poles. This discrepancy was ultimately resolved by Albert Einstein's special theory of relativity. According to the special theory of relativity, Newtonian mechanics breaks down when the relative speed between particles approaches the speed of light.

Coupled Oscillators

The solutions of this seemingly academic problem have far-reaching implications in many fields of physics. For example, a system of particles held together by springs turns out to be a useful model of the behaviour of atoms mutually bound in a crystalline solid.

To begin with a simple case, consider two particles in a line, as shown in figure. Each particle has mass m, each spring has spring constant k, and motion is restricted to the horizontal, or x, direction. Even this elementary system is capable of surprising behaviour, however. For instance, if one particle is held in place while the other is displaced, and then both are released, the displaced particle immediately begins to execute simple harmonic motion. This motion, by stretching the spring between the particles, starts to excite the second particle into motion. Gradually the energy of motion passes from the first particle to the second until a point is reached at which the first particle is at rest and only the second is oscillating. Then the process starts all over again, the energy passing in the opposite direction.

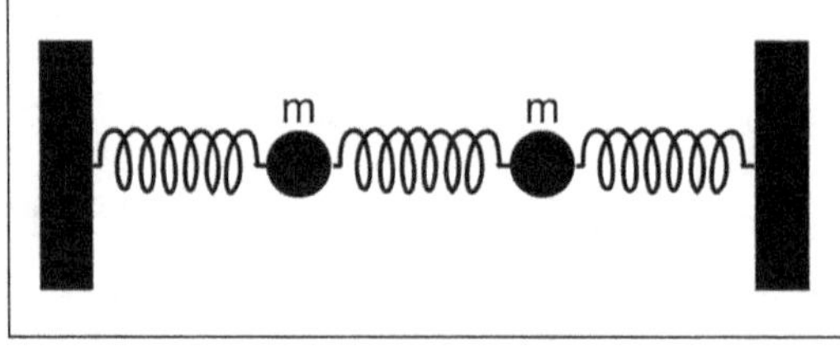

Coupled oscillators

To analyze the possible motions of the system, one writes equations similar to equation $a=\frac{d^2x}{dt^2}=-\frac{k}{m}x$, giving the acceleration of each particle owing to the forces acting on it. There is one equation for each particle (two equations in this case). The force on each particle depends not only on its displacement from its equilibrium position but also on its distance from the other particle, since the spring between them stretches or compresses according to that distance. For this reason the motions are coupled, the solution of each equation (the motion of each particle) depending on the solution of the other (the motion of the other).

Analyzing the system yields the fact that there are two special states of motion in which both particles are always in oscillation with the same frequency. In one state, the two particles oscillate in opposite directions with equal and opposite displacements from equilibrium at all times. In the other state, both particles move together, so that the spring between them is never stretched or compressed. The first of these motions has higher frequency than the second because the centre spring contributes an increase in the restoring force.

These two collective motions, at different, definite frequencies, are known as the normal modes of the system.

If a third particle is inserted into the system together with another spring, there will be three equations to solve, and the result will be three normal modes. A large number N of particles in a line will have N normal modes. Each normal mode has a definite frequency at which all the particles oscillate. In the highest frequency mode each particle moves in the direction opposite to both of its neighbours. In the lowest frequency mode, neighbours move almost together, barely disturbing the springs between them. Starting from one end, the amplitude of the motion gradually builds up, each particle moving a bit more than the one before, reaching a maximum at the centre, and then decreasing again. A plot of the amplitudes, shown in figure, basically describes one-half of a sine wave from one end of the system to the other. The next mode is a full sine wave, then $^3/_2$ of a sine wave, and so on to the highest frequency mode, which may be visualized as N/2 sine waves. If the vibrations were up and down rather than side to side, these modes would be identical to the fundamental and harmonic vibrations excited by plucking a guitar string.

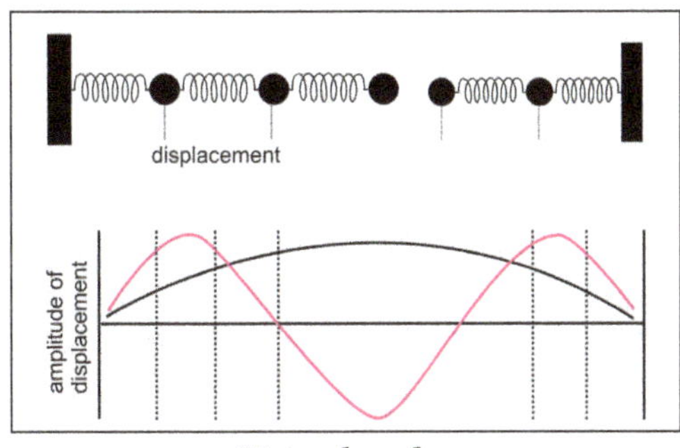

Normal modes

The atoms of a crystal are held in place by mutual forces of interaction that oppose any disturbance from equilibrium positions, just as the spring forces in the example above. For small displacements of the atoms, they behave mathematically just like spring forces—i.e., they obey Hooke's law, equation $F = -kx$. Each atom is free to move in three dimensions rather than one, however; therefore each atom added to a crystal adds three normal modes. In a typical crystal at ordinary temperature, all these modes are always excited by random thermal energy. The lower-frequency, longer-wavelength modes may also be excited mechanically. These are called sound waves.

Rigid Bodies

Statics

Statics is the study of bodies and structures that are in equilibrium. For a body to be in equilibrium, there must be no net force acting on it. In addition, there must be no net torque acting on it. Figure shows a body in equilibrium under the action of equal and opposite forces. Figure shows a body acted on by equal and opposite forces that produce a net torque, tending to start it rotating. It is therefore not in equilibrium.

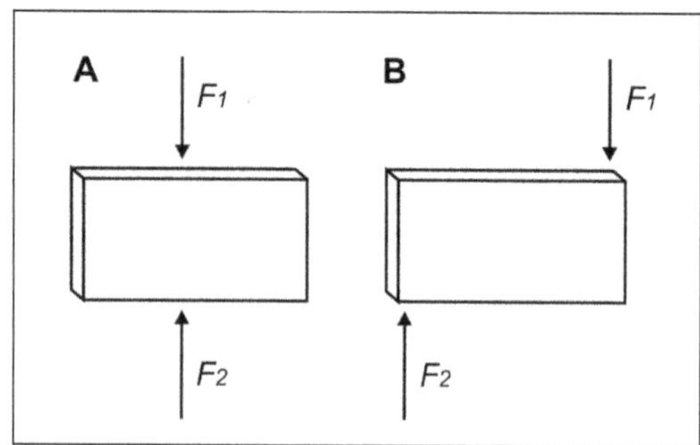

(A) A body in equilibrium under equal and opposite forces.
(B) A body not in equilibrium under equal and opposite forces.

When a body has a net force and a net torque acting on it owing to a combination of forces, all the forces acting on the body may be replaced by a single (imaginary) force called the resultant, which acts at a single point on the body, producing the same net force and the same net torque. The body can be brought into equilibrium by applying to it a real force at the same point, equal and opposite to the resultant. This force is called the equilibrant. An example is shown in figure.

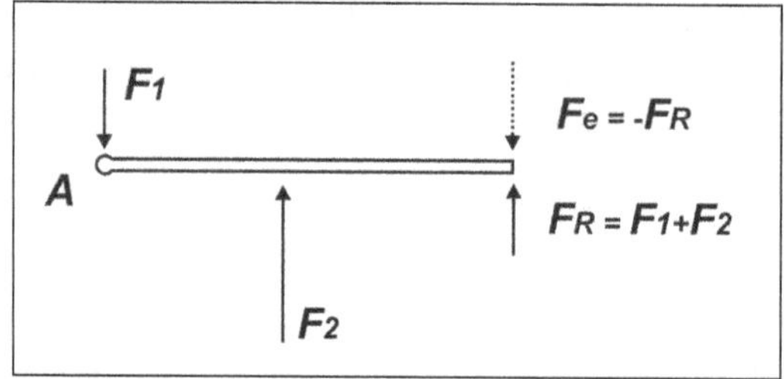

The resultant force (F_R) produces the same net force and the same net torque about point A as $F_1 + F_2$; the body can be brought into equilibrium by applying the equilibrant force F_e.

The torque on a body due to a given force depends on the reference point chosen, since the torque τ by definition equals r × F, where r is a vector from some chosen reference point to the point of application of the force. Thus, for a body to be at equilibrium, not only must the net force on it be equal to zero but the net torque with respect to any point must also be zero. Fortunately, it is easily shown for a rigid body that, if the net force is zero and the net torque is zero with respect to any one point, then the net torque is also zero with respect to any other point in the frame of reference.

A body is formally regarded as rigid if the distance between any set of two points in it is always constant. In reality no body is perfectly rigid. When equal and opposite forces are applied to a body, it is always deformed slightly. The body's own tendency to restore the deformation has the effect of applying counterforces to whatever is applying the forces, thus obeying Newton's third law. Calling a body rigid means that the changes in the dimensions of the body are small enough to be neglected, even though the force produced by the deformation may not be neglected.

Equal and opposite forces acting on a rigid body may act so as to compress the body or to stretch it. The bodies are then said to be under compression or under tension, respectively. Strings, chains, and cables are rigid under tension but may collapse under compression. On the other hand, certain building materials, such as brick and mortar, stone, or concrete, tend to be strong under compression but very weak under tension.

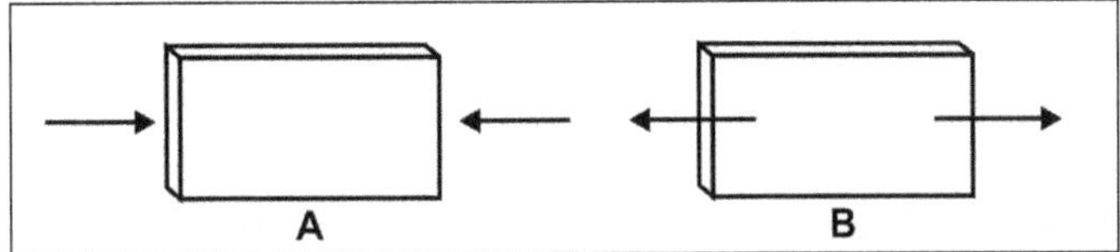

(A) Compression produced by equal and opposite forces. (B) Tension produced by equal and opposite forces.

The most important application of statics is to study the stability of structures, such as edifices and bridges. In these cases, gravity applies a force to each component of the structure as well as to any bodies the structure may need to support. The force of gravity acts on each bit of mass of which each component is made, but for each rigid component it may be thought of as acting at a single point, the centre of gravity, which is in these cases the same as the centre of mass.

To give a simple but important example of the application of statics, consider the two situations shown in figure. In each case, a mass m is supported by two symmetric members, each making an angle θ with respect to the horizontal. In figure the members are under tension; in figure they are under compression. In either case, the force acting along each of the members is shown to be,

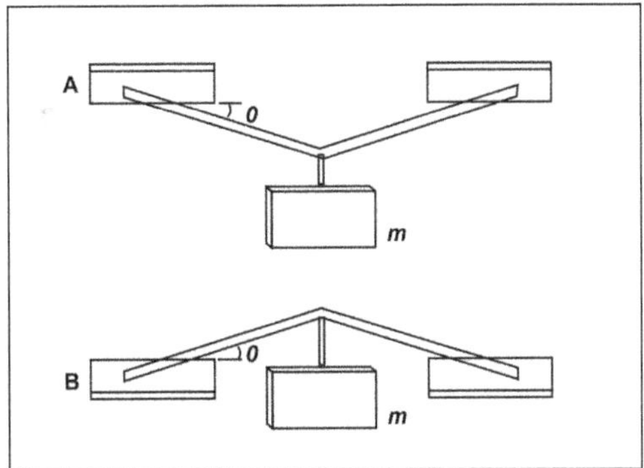

(A) A body supported by two rigid members under tension.
(B) A body supported by two rigid members under compression.

$$F = \frac{mg}{2\sin\theta}.$$

The force in either case thus becomes intolerably large if the angle θ is allowed to be very small. In other words, the mass cannot be hung from thin horizontal members only capable of carrying either the compression or the tension forces of the mass.

The ancient Greeks built magnificent stone temples; however, the horizontal stone slabs that constituted the roofs of the temples could not support even their own weight over more than a very small span. For this reason, one characteristic that identifies a Greek temple is the many closely spaced pillars needed to hold up the flat roof. The problem posed by equation $F = \frac{mg}{2\sin\theta}$ was solved by the ancient Romans, who incorporated into their architecture the arch, a structure that supports its weight by compression, corresponding to figure.

A suspension bridge illustrates the use of tension. The weight of the span and any traffic on it is supported by cables, which are placed under tension by the weight. Corresponding to figure, the cables are not stretched to be horizontal, but rather they are always hung so as to have substantial curvature.

It should be mentioned in passing that equilibrium under static forces is not sufficient to guarantee the stability of a structure. It must also be stable against perturbations such as the additional forces that might be imposed, for example, by winds or by earthquakes. Analysis of the stability of structures under such perturbations is an important part of the job of an engineer or architect.

Rotation about a Fixed Axis

Consider a rigid body that is free to rotate about an axis fixed in space. Because of the body's inertia, it resists being set into rotational motion, and equally important, once rotating, it resists being brought to rest.

Take the axis of rotation to be the z-axis. A vector in the x-y plane from the axis to a bit of mass fixed in the body makes an angle θ with respect to the x-axis. If the body is rotating, θ changes with time, and the body's angular frequency is:

$$\omega = \frac{d\theta}{dt};$$

ω is also known as the angular velocity. If ω is changing in time, there is also an angular acceleration α, such that,

$$\alpha = \frac{d\omega}{dt}.$$

Because linear momentum p is related to linear speed v by p = mv, where m is the mass, and because force F is related to acceleration a by F = ma, it is reasonable to assume that there exists a quantity I that expresses the rotational inertia of the rigid body in analogy to the way m expresses the inertial resistance to changes in linear motion. One would expect to find that the angular momentum is given by:

$$L = l\omega$$

and that the torque (twisting force) is given by:

$$\tau = I\alpha$$

One can imagine dividing the rigid body into bits of mass labeled m_1, m_2, m_3, and so on. Let the bit of mass at the tip of the vector be called mi, as indicated in figure. If the length of the vector from the axis to this bit of mass is Ri, then mi's linear velocity vi equals ωRi, and its angular momentum Li equals miviRi, or $miRi^2\omega$. The angular momentum of the rigid body is found by summing all the contributions from all the bits of mass labeled i = 1, 2, 3 . . . :

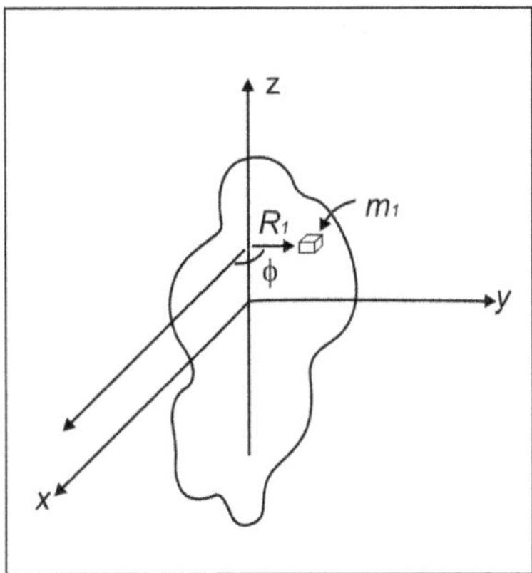

Rotation around a fixed axis.

$$L = \left(\sum_i m_i R_i^2 \right) \omega.$$

In a rigid body, the quantity in parentheses in equation $L = \left(\sum_i m_i R_i^2 \right) \omega$ is always constant (each bit of mass m_i always remains the same distance R_i from the axis). Thus if the motion is accelerated, then:

$$\frac{dL}{dt} = \left(\sum_i m_i R_i^2 \right) \frac{d\omega}{dt}.$$

Recalling that τ = dL/dt, one may write,

$$\tau = \left(\sum_i m_i R_i^2 \right) \alpha.$$

(These equations may be written in scalar form, since L and τ are always directed along the axis of rotation in this discussion.) Comparing equations $L = \left(\sum_i m_i R_i^2 \right) \omega$ and $\tau = \left(\sum_i m_i R_i^2 \right) \alpha$ with $L = I\omega$ and $\tau = I\alpha$, one finds that:

$$I = \sum_i m_i R_i^2.$$

The quantity I is called the moment of inertia.

According to equation $I = \sum_i m_i R_i^2$, the effect of a bit of mass on the moment of inertia depends on its distance from the axis. Because of the factor Ri^2, mass far from the axis makes a bigger contribution than mass close to the axis. It is important to note that Ri is the distance from the axis, not from a point. Thus, if xi and yi are the x and y coordinates of the mass mi, then $Ri^2 = xi^2 + yi^2$, regardless of the value of the z coordinate. The moments of inertia of some simple uniform bodies are given in the table.

Moments of Interia for uniform Bodies		
Body	Axis	L
Thin rod (lengthb L)	Perpendicular axis through centre	$\frac{1}{12}ML^2$
Thin ring (radius R)	Perpendicular axis through centre	MR^2
Solid circular cylinder	Axis of cylinder	$\frac{1}{2}MR^2$
Thin desk	Transverse axis through centre	$\frac{1}{4}MR^2$
Solid sphere	Any axis through centre	$\frac{2}{5}MR^2$
Thin spherical cell	Any axis through centre	$\frac{2}{3}MR^2$
Rectangular plate length a height b	Axis through centre perpendicular of the plate	$\frac{1}{12}M(a^2+b^2)$

The moment of inertia of any body depends on the axis of rotation. Depending on the symmetry of the body, there may be as many as three different moments of inertia about mutually perpendicular axes passing through the centre of mass. If the axis does not pass through the centre of mass, the moment of inertia may be related to that about a parallel axis that does so. Let Ic be the moment of inertia about the parallel axis through the centre of mass, r the distance between the two axes, and M the total mass of the body. Then,

$$I = I_c + Mr^2,$$

In other words, the moment of inertia about an axis that does not pass through the centre of mass is equal to the moment of inertia for rotation about an axis through the centre of mass (I_c) plus a contribution that acts as if the mass were concentrated at the centre of mass, which then rotates about the axis of rotation.

The dynamics of rigid bodies rotating about fixed axes may be summarized in three equations. The angular momentum is $L = I\omega$, the torque is $\tau = I\alpha$, and the kinetic energy is $K = {}^1/_2 I\omega^2$.

Rotation about a Moving Axis

The general motion of a rigid body tumbling through space may be described as a combination of translation of the body's centre of mass and rotation about an axis through the centre of mass. The linear momentum of the body of mass M is given by:

$$P = Mv_c,$$

where v_c is the velocity of the centre of mass. Any change in the momentum is governed by Newton's second law, which states that:

$$F = \frac{dp}{dt},$$

where F is the net force acting on the body. The angular momentum of the body with respect to any reference point may be written as:

$$L = L_c + r \times p,$$

where L_c is the angular momentum of rotation about an axis through the centre of mass, r is a vector from the reference point to the centre of mass, and r × p is therefore the angular momentum associated with motion of the centre of mass, acting as if all the body's mass were concentrated at that point. The quantity L_c in equation $L = L_c + r \times p$ is sometimes called the body's spin, and r × p is called the orbital angular momentum. Any change in the angular momentum of the body is given by the torque equation,

$$\tau = \frac{dL}{dt}.$$

An example of a body that undergoes both translational and rotational motion is the Earth, which rotates about an axis through its centre once per day while executing an orbit around the Sun once per year. Because the Sun exerts no torque on the Earth with respect to its own centre, the orbital angular momentum of the Earth is constant in time. However, the Sun does exert a small torque on the Earth with respect to the planet's centre, owing to the fact that the Earth is not perfectly spherical. The result is a slow shifting of the Earth's axis of rotation, known as the precession of the equinoxes.

The kinetic energy of a body that is both translating and rotating is given by:

$$K = \frac{1}{2} Mv_c^2 + \frac{1}{2} I\omega^2,$$

where I is the moment of inertia and ω is the angular velocity of rotation about the axis through the centre of mass.

A common example of combined rotation and translation is rolling motion, as exhibited by a billiard ball rolling on a table, or a ball or cylinder rolling down an inclined plane. Consider the latter example, illustrated in figure. Motion is impelled by the force of gravity, which may be resolved into two components, FN, which is normal to the plane, and Fp, which is parallel to it. In addition to gravity, friction plays an essential role. The force of friction, written as f, acts parallel to the plane, in opposition to the direction of motion, at the point of contact between the plane and the rolling body. If f is very small, the body will slide without rolling. If f is very large, it will prevent motion from occurring. The magnitude of f depends on the smoothness and composition of the body and the plane, and it is proportional to FN, the normal component of the force.

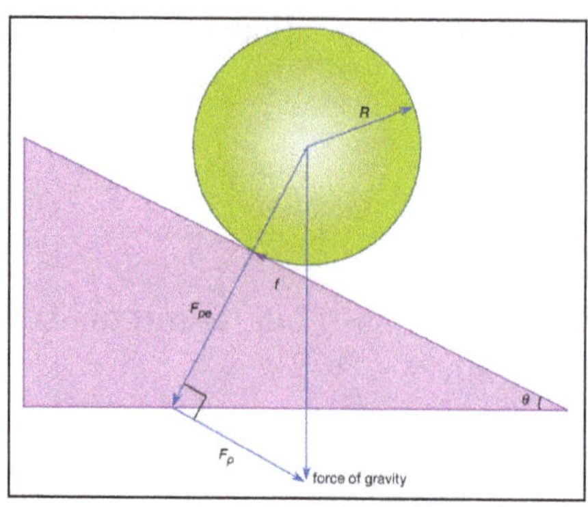

Rolling motion

Consider a case in which f is just large enough to cause the body (sphere or cylinder) to roll without slipping. The motion may be analyzed from the point of view of an axis passing through the point of contact between the rolling body and the plane. Remarkably, the point of contact may always be regarded to be instantaneously at rest. To understand why, suppose that the rolling body has radius R and angular velocity ω about its centre-of-mass axis. Then, with respect to its own axis, each point on the circular cross section in figure. moves with instantaneous tangential linear speed $v_c = R\omega$. In particular, the point of contact is moving backward with this speed relative to the centre of mass. But with respect to the inclined plane, the centre of mass is moving forward with exactly this same speed. The net effect of the two equal and opposite speeds is that the point of contact is always instantaneously at rest. Therefore, although friction acts at that point, no work is done by friction, so mechanical energy (potential plus kinetic) may be regarded as conserved.

With respect to the axis through the point of contact, the torque is equal to RF_p, giving rise to an angular acceleration α given by $I_p\alpha = RF_p$, where I_p is the moment of inertia about the point-of-contact axis and can be determined by applying equation $I = \sum_i m_i R_i^2$ relating moments of inertia about parallel axes ($I_p = I + MR^2$). Thus,

$$\alpha = \frac{RF_p}{1 + MR^2}.$$

From this result, the motion of the body is easily obtained using the fact that the velocity of the centre of mass is $v_c = R\omega$ and hence the linear acceleration of the centre of mass is $a_c = R\alpha$.

Notice that, although without friction no angular acceleration would occur, the force of friction does not affect the magnitude of α. Because friction does no work, this same result may be obtained by applying energy conservation. The situation also may be analyzed entirely from the point of view of the centre of mass. In that case, the torque is −fR, but f also provides a linear force on the body. The f may then be eliminated by using Newton's second law and the fact that the torque equals the moment of inertia times the angular acceleration, once again leading to the same result.

One more interesting fact is hidden in the form of equation $\alpha = \frac{RF_p}{1 + MR^2}$. The parallel component of the force of gravity is given by:

$$F_p = Mg\sin\theta,$$

Where θ is the angle of inclination of the plane. The moment of inertia about the centre of mass of any body of mass M may be written:

$$I = Mk^2,$$

where k is a distance called the radius of gyration. Comparison to equation $I = \sum_i m_i R_i^2$ shows that k is a measure of how far from the centre of mass the mass of the body is concentrated. Using equations $F_p = Mg\sin\theta$, and $I = Mk^2$, in equation $\alpha = \frac{RF_p}{1 + MR^2}$, one finds that:

$$\alpha = \frac{Rg\sin\theta}{k^2 + R^2}.$$

Thus, the angular acceleration of a body rolling down a plane does not depend on its total mass, although it does depend on its shape and distribution of mass. The same may be said of a_c, the linear acceleration of the centre of mass. The acceleration of a rolling ball, like the acceleration of a freely falling object, is independent of its mass. This observation helps to explain why Galileo was able to discover many of the basic laws of dynamics in gravity by studying the behaviour of balls rolling down inclined planes.

Motion in a Rotating Frame

Centrifugal Force

According to the principle of Galilean relativity, if Newton's laws are true in any reference frame, they are also true in any other frame moving at constant velocity with respect to the first one. Conversely, they do not appear to be true in any frame accelerated with respect to the first. Instead, in an accelerated frame, objects appear to have forces acting on them that are not in fact present. These are called pseudoforces. Since rotational motion is always accelerated motion, pseudoforces may always be observed in rotating frames of reference.

As one example, a frame of reference in which the Earth is at rest must rotate once per year about the Sun. In this reference frame, the gravitational force attracting the Earth toward the Sun appears to be balanced by an equal and opposite outward force that keeps the Earth in stationary equilibrium. This outward pseudoforce, is the centrifugal force.

The rotation of the Earth about its own axis also causes pseudoforces for observers at rest on the Earth's surface. There is a centrifugal force, but it is much smaller than the force of gravity. Its effect is that, at the Equator, where it is largest, the gravitational acceleration g is about 0.5 percent smaller than at the poles, where there is no centrifugal force. This same centrifugal force is responsible for the fact that the Earth is slightly nonspherical, bulging just a bit at the Equator.

Pseudoforces can have real consequences. The oceanic tides on Earth, for example, are a consequence of centrifugal forces in the Earth-Moon and Earth-Sun systems. The Moon appears to be orbiting the Earth, but in reality both the Moon and the Earth orbit their common centre of mass. The centre of mass of the Earth-Moon system is located inside the Earth nearly three-fourths of the distance from the centre to the surface, or roughly 4,700 kilometres from the centre of the Earth. The Earth rotates about this point approximately once a month. The gravitational attraction of the Moon and the centrifugal force of this rotation are exactly balanced at the centre of the Earth. At the surface of the Earth closest to the Moon, the Moon's gravity is stronger than the centrifugal force. The ocean's waters, which are free to move in response to this unbalanced force, tend to build up a small bulge at that point. On the surface of the Earth exactly opposite the Moon, the centrifugal force is stronger than the Moon's gravity, and a small bulge of water tends to build up there as well. The water is correspondingly depleted at the points 90° on either side of these. Each day the Earth rotates beneath these bulges and troughs, which remain stationary with respect to the Earth-Moon system. The result is two high tides and two low tides every day every place on Earth. The Sun has a similar effect, but of only about half the size; it increases or decreases the size of the tides depending on its relative alignment with the Earth and Moon.

Coriolis Force

The Coriolis force is a pseudoforce that operates in all rotating frames. One way to envision it is to imagine a rotating platform (such as a merry-go-round or a phonograph turntable) with a perfectly smooth surface and a smooth block sliding inertially across it. The block, having no (real) forces acting on it, moves in a straight line at constant speed in inertial space. However, the platform rotates under it, so that to an observer on the platform, the block appears to follow a curved trajectory, bending in the opposite direction to the motion of the platform. Since the motion is curved, and hence accelerated, there appears, to the observer, to be a force operating. That pseudoforce is called the Coriolis force.

The Coriolis force also may be observed on the surface of the Earth. For example, many science museums have a pendulum, called a Foucault pendulum, suspended from a long cable with markers to show that its plane of motion rotates slowly. The rotation of the plane of motion is caused by the Coriolis force. The effect is most easily imagined by picturing the pendulum swinging directly above the North Pole. The plane of its motion remains stationary in inertial space, while the Earth rotates once a day beneath it.

At lower latitudes, the effect is a bit more subtle, but it is still present. Imagine that, somewhere in the Northern Hemisphere, a projectile is fired due south. As viewed from inertial space, the projectile initially has an eastward component of velocity as well as a southward component because the gun that fired it, which is stationary on the surface of the Earth, was moving eastward with the Earth's rotation at the instant it was fired. However, since it was fired to the south, it lands at a slightly lower latitude, closer to the Equator. As one moves south, toward the Equator, the tangential speed of the Earth's surface due to its rotation increases because the surface is farther from the axis of rotation. Thus, although the projectile has an eastward component of velocity (in inertial space), it lands at a place where the surface of the Earth has a larger eastward component of velocity. Thus, to the observer on Earth, the projectile seems to curve slightly to the west. That westward curve is attributed to the Coriolis force. If the projectile were fired to the north, it would seem to curve eastward.

The same analysis applied to a Foucault pendulum explains why its plane of motion tends to rotate in the clockwise direction anywhere in the Northern Hemisphere and in the counterclockwise direction in the Southern Hemisphere. Storms, known as cyclones, tend to rotate in the opposite direction in each hemisphere, also due to the Coriolis force. Air moves in all directions toward a low-pressure centre. In the Northern Hemisphere, air moving up from the south is deflected eastward, while air moving down from the north is deflected westward. This effect tends to give cyclones a counterclockwise circulation in the Northern Hemisphere. In the Southern Hemisphere, cyclones tend to circulate in the clockwise direction.

Spinning Tops and Gyroscopes

Figure shows a wheel that is weighted in its rim to maximize its moment of inertia I and that is spinning with angular frequency ω on a horizontal axle supported at both ends. As shown, it has an angular momentum L along the x direction equal to Iω. Now suppose the support at point P is removed, leaving the axle supported only at one end. Gravity, acting on the mass of the wheel as if it were concentrated at the centre of mass, applies a downward force on the wheel. The wheel, however, does not fall. Instead, the axle remains (nearly) horizontal but rotates in the counterclockwise direction as seen from above. This motion is called gyroscopic precession.

Horizontal precession occurs in this case because the gravitational force results in a torque with respect to the point of suspension, such that $\boldsymbol{\tau} = r \times F$ and is directed, initially, in the positive y direction. The torque causes the angular momentum L to move toward that direction according to $\boldsymbol{\tau} = dL/dt$. Because $\boldsymbol{\tau}$ is perpendicular to L, it does not change the magnitude of the angular momentum, only its direction. As precession proceeds, the torque remains horizontal, and the angular momentum vector, continually redirected by the torque, executes uniform circular motion in the horizontal plane at a frequency Ω, the frequency of precession.

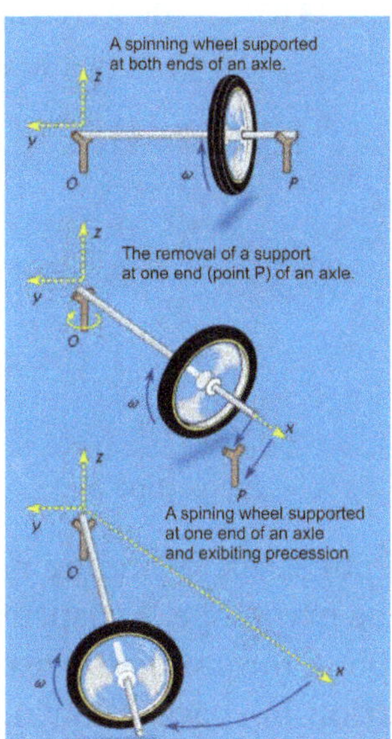

Gyroscopic precession

In reality, the motion is a bit more complicated than uniform precession in the horizontal plane. When the support at P is released, the centre of mass of the wheel initially drops slightly below the horizontal plane. This drop reduces the gravitational potential energy of the system, releasing kinetic energy for the orbital motion of the centre of mass as it precesses. It also provides a small component of L in the negative z direction, which balances the angular momentum in the positive z direction that results from the orbital motion of the centre of mass. There can be no net angular momentum in the vertical direction because there is no component of torque in that direction.

One more complication: the initial drop of the centre of mass carries it too far for a stable plane of precession, and it tends to bounce back up after overshooting. This produces an up-and-down oscillation during precession, called nutation ("nodding"). In most cases, nutation is quickly damped by friction in the bearings, leaving uniform precession.

A spinning top undergoes all the motions described above. If it is initially set spinning with a vertical axis, there will be virtually no torque, and conservation of angular momentum will keep the axis vertical for a long time. Eventually, however, friction at the point of contact will require the centre of mass to lower itself, which can only happen if the axis tilts. The spinning will also slow down, making the tilting process easier. Once the top tilts, gravity produces a horizontal torque that leads to precession of the spin axis. The subsequent motion depends on whether the point of contact is fixed or free to slip on the horizontal plane. Vast tomes have been written on the motions of tops.

A gyroscope is a device that is designed to resist changes in the direction of its axis of spin. That purpose is generally accomplished by maximizing its moment of inertia about the spin axis and by spinning it at the maximum practical frequency. Each of these considerations has the effect of maximizing the magnitude of the angular momentum, thus requiring a larger torque to change its direction. It is quite generally true that the torque τ, the angular momentum L, and the precession frequency Ω (defined as a vector along the precession axis in the direction given by the right-hand rule) are related by:

$$\tau = \Omega \times L.$$

Equation $\tau = \Omega \times L$, illustrated in figure, is called the gyroscope equation.

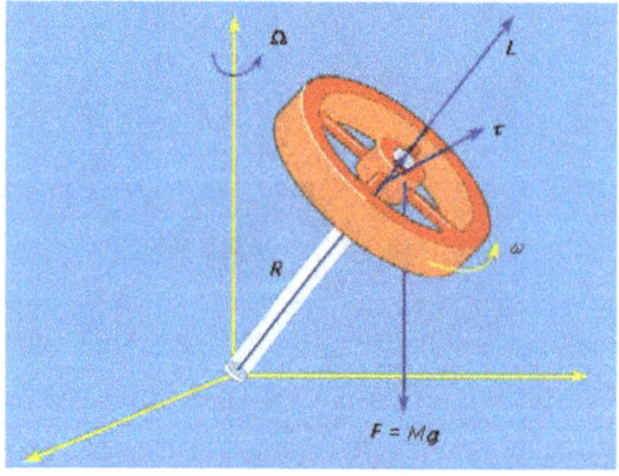

A gyroscope

Gyroscopes are used for a variety of purposes, including navigation. Use of gyroscopes for this purpose is called inertial guidance. The gyroscope is suspended as nearly as possible at its centre of mass, so that gravity does not apply a torque that causes it to precess. The gyroscope tends therefore to point in a constant direction in space, allowing the orientation of the vehicle to be accurately maintained.

One further application of the gyroscope principle may be seen in the precession of the equinoxes. The Earth is a kind of gyroscope, spinning on its axis once each day. The Sun would apply no torque to the Earth if the Earth were perfectly spherical, but it is not. The Earth bulges slightly at the Equator. As indicated in figure, the effect of the Sun's gravity on the near bulge (larger than it is on the far bulge) results in a net torque about the centre of the Earth. When the Earth is on the other side of the Sun, the net torque remains in the same direction. The torque is small but persistent. It causes the axis of the Earth to precess, about one revolution every 25,800 years.

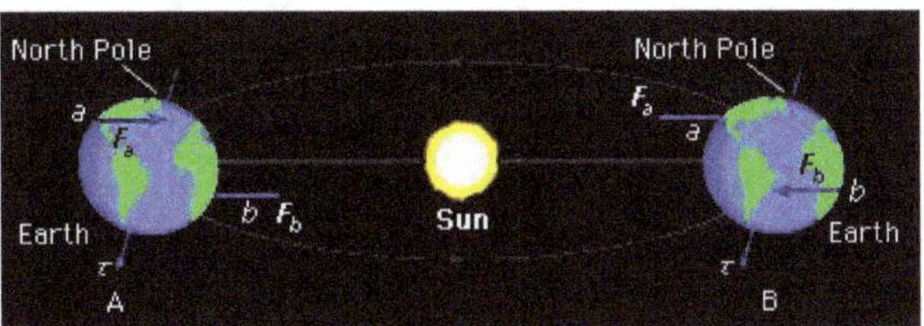

Forces acting on equatorial bulges in (A) the summer and (B) the winter cause the axis of the Earth to precess.

As seen from the Earth, the Sun passes through the plane of the Equator twice each year. These points are called the equinoxes, and on the days of the equinoxes the hours of daylight and night are equal. From antiquity it has been known that the point in the sky where the Sun intersects the plane of the Equator is not the same each year but rather drifts very slowly to the west. This ancient observation, first explained by Newton, is due to the precession of the Earth's axis. It is called the precession of the equinoxes.

Analytic Approaches

Classical mechanics can, in essence, be reduced to Newton's laws, starting with the second law, in the form:

$$F = \frac{dp}{dt}.$$

If the net force acting on a particle is F, knowledge of F permits the momentum p to be found; and knowledge of p permits the position r to be found, by solving the equation:

$$\frac{dr}{dt} = \frac{p}{m}.$$

These solutions give the components of p—that is, p_x, p_y, and p_z—and the components of r—x, y, and z—each as a function of time. To complete the solution, the value of each quantity—p_x, p_y, p_z, x, y, and z—must be known at some definite time, say, t = 0. If there is more than one particle, an equation in the form of equation $F = \frac{dp}{dt}$ must be written for each particle, and the solution will involve finding the six variables x, y, z, p_x, p_y, and p_z, for each particle as a function of time, each once again subject to some initial condition. The equations may not be independent, however. For example, if the particles interact with one another, the forces will be related by Newton's third law. In this case (and others), the forces may also depend on time.

If the problem involves more than a very few particles, this method of solution quickly becomes intractable. Furthermore, in many cases it is not useful to express the problem purely in terms of particles and forces. Consider, for example, the problem of a sphere or cylinder rolling without slipping on a plane surface. Rolling without slipping is produced by friction due to forces acting between atoms in the rolling body and atoms in the plane, but the interactions are very complex; they probably are not fully understood even today, and one would like to be able to formulate and solve the problem without introducing them or needing to understand them. For all these reasons, methods that go beyond solving equations $F = \frac{dp}{dt}$ and $\frac{dr}{dt} = \frac{p}{m}$ have had to be introduced into classical mechanics.

The methods that have been introduced do not involve new physics. In fact, they are deduced directly from Newton's laws. They do, however, involve new concepts, new language to describe those concepts, and the adoption of powerful mathematical techniques.

Configuration Space

The position of a single particle is specified by giving its three coordinates, x, y, and z. To specify the positions of two particles, six coordinates are needed, x_1, y_1, z_1, x_2, y_2, z_2. If there are N particles, 3N coordinates will be needed. Imagine a system of 3N mutually orthogonal coordinates in a 3N-dimensional space (a space of more than three dimensions is a purely mathematical construction, sometimes known as a hyperspace). To specify the exact position of one single point in this space, 3N coordinates are needed. However, one single point can represent the entire configuration of all N particles in the problem. Furthermore, the path of that single point as a function of time is the complete solution of the problem. This 3N-dimensional space is called configuration space.

Configuration space is particularly useful for describing what is known as constraints on a problem. Constraints are generally ways of describing the effects of forces that are best not explicitly introduced into the problem. For example, consider the simple case of a falling body near the surface of the Earth. The equations of motion—equations $z = z_0 - \frac{1}{2}gt^2$, $v \quad gt$, and $a = g$ —are valid only until the body hits the ground. Physically, this restriction is due to forces between atoms in the falling body and atoms in the ground, but, as a practical matter, it is preferable to say that the solutions are valid only for $z > 0$ (where $z = 0$ is ground level). This constraint, in the form of an inequality, is very difficult to incorporate directly into the equations of the problem. In the language of configuration space, however, one merely needs to specify that the problem is being solved only in the region of configuration space for which $z > 0$.

Notice that the constraint mentioned above, rolling without sliding on a plane, cannot easily be described in configuration space, since it is basically a condition on relative velocities of rotation and translation; but another constraint, that the body is restricted to motion along the plane, is easily described in configuration space.

Another type of constraint specifies that a body is rigid. Then, even though the body is composed of a very large number of atoms, it is not necessary to find separately the x, y, and z coordinate of each atom because these are related to those of the other atoms by the condition of rigidity. A careful analysis yields that, rather than needing 3N coordinates (where N may be, for example, 10^{24} atoms), only 6 are needed: 3 to specify the position of the centre of mass and 3 to give the orientation of the body. Thus, in this case, the constraint has reduced the number of independent coordinates from 3N to 6. Rather than restricting the behaviour of the system to a portion of the original 3N-dimensional configuration space, it is possible to describe the system in a much simpler 6-dimensional configuration space. It should be noted, however, that the six coordinates are not necessarily all distances. In fact, the most convenient coordinates are three distances (the x, y, and z coordinates of the centre of mass of the body) and three angles, which specify the orientation of a set of axes fixed in the body relative to a set of axes fixed in space. This is an example of the use of constraints to reduce the number of dynamic variables in a problem (the x, y, and z coordinates of each particle) to a smaller number of generalized dynamic variables, which need not even have the same dimensions as the original ones.

The Principle of Virtual Work

A special class of problems in mechanics involves systems in equilibrium. The problem is to find the configuration of the system, subject to whatever constraints there may be, when all forces are balanced. The body or system will be at rest (in the inertial rest frame of its centre of mass), meaning that it occupies one point in configuration space for all time. The problem is to find that point. One criterion for finding that point, which makes use of the calculus of variations, is called the principle of virtual work.

According to the principle of virtual work, any infinitesimal virtual displacement in configuration space, consistent with the constraints, requires no work. A virtual displacement means an instantaneous change in coordinates (a real displacement would require finite time during which particles might move and forces might change). To express the principle, label the generalized coordinates $r_1, r_2, \ldots, r_i, \ldots$. Then if F_i is the net component of generalized force acting along the coordinate ri,

$$\sum_i F_i dr_i = 0.$$

Here, $F_i\,dr_i$ is the work done when the generalized coordinate is changed by the infinitesimal amount dri. If ri is a real coordinate (say, the x coordinate of a particle), then F_i is a real force. If r_i is a generalized coordinate (say, an angular displacement of a rigid body), then F_i is the generalized force such that $F_i\,dr_i$ is the work done (for an angular displacement, F_i is a component of torque).

Take two simple examples to illustrate the principle. First consider two particles that are restricted to motion in the x direction and are constrained by a taut string connecting them. If their x coordinates are called x_1 and x_2, then $F_1dx_1 + F_2dx_2 = 0$ according to the principle of virtual work. But the taut string requires that the particles be displaced the same amount, so that $dx_1 = dx_2$, with the result that $F_1 + F_2 = 0$. The particles might be in equilibrium, for example, under equal and opposite forces, but F_1 and F_2 do not need individually to be zero. This is generally true of the Fi in equation $\sum_i F_i dr_i = 0$. As a second example, consider a rigid body in space. Here, the constraint of rigidity has already been expressed by reducing the coordinate space to that of six generalized coordinates. These six coordinates (x, y, z, and three angles) can change quite independently of one another. In other words, in equation $\sum_i F_i dr_i = 0$, the six dri are arbitrary. Thus, the only way equation $\sum_i F_i dr_i = 0$ can be satisfied is if all six Fi are zero. This means that the rigid body can have no net component of force and no net component of torque acting on it. Of course, this same conclusion was reached earlier by less abstract arguments.

Lagrange's and Hamilton's Equations

Elegant and powerful methods have also been devised for solving dynamic problems with constraints. One of the best known is called Lagrange's equations. The Lagrangian L is defined as L = T – V, where T is the kinetic energy and V the potential energy of the system in question. Generally speaking, the potential energy of a system depends on the coordinates of all its particles; this may be written as $V = V(x_1, y_1, z_1, x_2, y_2, z_2, \ldots)$. The kinetic energy generally depends on the velocities, which, using the notation $vx = d_x/dt = \dot{x}$, may be written $T = T(\dot{x}_1, \dot{y}_1, \dot{z}_1, \dot{x}_2, \dot{y}_2, \dot{z}_2, \ldots)$. Thus, a dynamic problem has six dynamic variables for each particle—that is, x, y, z and $\dot{x}$, $\dot{y}$, $\dot{z}$—and the Lagrangian depends on all 6N variables if there are N particles.

In many problems, however, the constraints of the problem permit equations to be written relating at least some of these variables. In these cases, the 6N related dynamic variables may be reduced to a smaller number of independent generalized coordinates (written symbolically as q_1, $q_2, \ldots qi, \ldots$) and generalized velocities (written as $\dot{q}_1, \dot{q}_2, \ldots \dot{q}i, \ldots$), just as, for the rigid body, 3N coordinates were reduced to six independent generalized coordinates (each of which has an associated velocity). The Lagrangian, then, may be expressed as a function of all the qi and $\dot{q}$i. It is possible, starting from Newton's laws only, to derive Lagrange's equations:

$$\frac{d}{dt}\frac{\partial L}{\partial \dot{q}_i} - \frac{\partial L}{\partial q_i} = 0;$$

where the notation ∂L/∂qi means differentiate L with respect to qi only, holding all other variables constant. There is one equation of the form $\frac{d}{dt}\frac{\partial L}{\partial \dot{q}_i} - \frac{\partial L}{\partial q_i} = 0$ for each of the generalized coordinates qi (e.g., six equations for a rigid body), and their solutions yield the complete dynamics of the system. The use of generalized coordinates allows many coupled equations of the form $F = \frac{dp}{dt}$ to be reduced to fewer, independent equations of the form $\frac{d}{dt}\frac{\partial L}{\partial \dot{q}_i} - \frac{\partial L}{\partial q_i} = 0$.

There is an even more powerful method called Hamilton's equations. It begins by defining a generalized momentum p_i, which is related to the Lagrangian and the generalized velocity $\dot{q}_i$ by $p_i = \partial L/\partial \dot{q}_i$. A new function, the Hamiltonian, is then defined by $H = \Sigma i\, \dot{q}_i p_i - L$. From this point it is not difficult to derive:

$$\dot{q}_i = \frac{\partial H}{\partial p_i}$$

And

$$-\dot{p}_i = \frac{\partial H}{\partial q_i}.$$

These are called Hamilton's equations. There are two of them for each generalized coordinate. They may be used in place of Lagrange's equations, with the advantage that only first derivatives—not second derivatives—are involved.

The Hamiltonian method is particularly important because of its utility in formulating quantum mechanics. However, it is also significant in classical mechanics. If the constraints in the problem do not depend explicitly on time, then it may be shown that H = T + V, where T is the kinetic energy and V is the potential energy of the system—i.e., the Hamiltonian is equal to the total energy of the system. Furthermore, if the problem is isotropic (H does not depend on direction in space) and homogeneous (H does not change with uniform translation in space), then Hamilton's equations immediately yield the laws of conservation of angular momentum and linear momentum, respectively.

Kinematics

Kinematics is the study of the motion of points, objects, and groups of objects without considering the causes of its motion.

Kinematics is the branch of classical mechanics that describes the motion of points, objects and systems of groups of objects, without reference to the causes of motion (i.e., forces). The study of kinematics is often referred to as the "geometry of motion."

Objects are in motion all around us. Everything from a tennis match to a space-probe flyby of the planet Neptune involves motion. When you are resting, your heart moves blood through your veins. Even in inanimate objects there is continuous motion in the vibrations of atoms and molecules. Interesting questions about motion can arise: how long will it take for a space probe to travel to Mars? Where will a football land if thrown at a certain angle? An understanding of motion, however, is also key to understanding other concepts in physics. An understanding of acceleration, for example, is crucial to the study of force.

To describe motion, kinematics studies the trajectories of points, lines and other geometric objects, as well as their differential properties (such as velocity and acceleration). Kinematics is used in astrophysics to describe the motion of celestial bodies and systems; and in mechanical engineering, robotics and biomechanics to describe the motion of systems composed of joined parts (such as an engine, a robotic arm, or the skeleton of the human body).

A formal study of physics begins with kinematics.Kinematic analysis is the process of measuring the kinematic quantities used to describe motion. The study of kinematics can be abstracted into purely mathematical expressions, which can be used to calculate various aspects of motion such as velocity, acceleration, displacement, time, and trajectory.

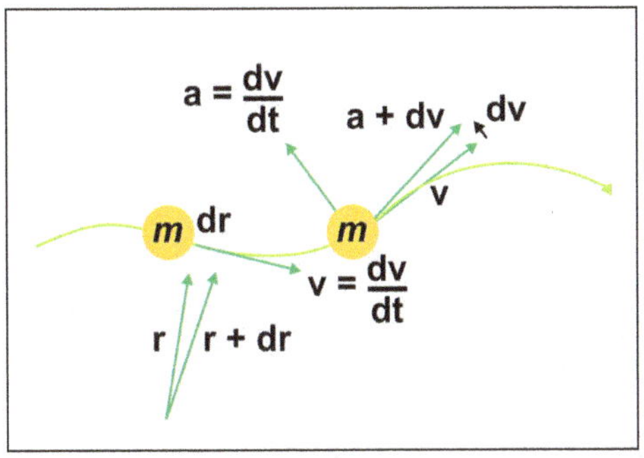

Kinematics of a particle trajectory: Kinematic equations can be used to calculate the trajectory of particles or objects. The physical quantities relevant to the motion of a particle include: mass m, position r, velocity v, acceleration a.

Reference Frames and Displacement

In order to describe an object's motion, you need to specify its position relative to a convenient reference frame.

In order to describe the motion of an object, you must first describe its position — where it is at any particular time. More precisely, you need to specify its position relative to a convenient reference frame. Earth is often used as a reference frame, and we often describe the position of objects related to its position to or from Earth. Mathematically, the position of an object is generally represented by the variable x.

Frames of Reference

There are two choices you have to make in order to define a position variable x. You have to decide where to put x = 0 and which direction will be positive. This is referred to as choosing a coordinate system, or choosing a frame of reference. As long as you are consistent, any frame is equally valid. But you don't want to change coordinate systems in the middle of a calculation. Imagine sitting in a train in a station when suddenly you notice that the station is moving backward. Most people would say that they just failed to notice that the train was moving — it only seemed like the station was moving. But this shows that there is a third arbitrary choice that goes into choosing a coordinate system: valid frames of reference can differ from each other by moving relative to one another. It might seem strange to use a coordinate system moving relative to the earth — but, for instance, the frame of reference moving along with a train might be far more convenient for describing things happening inside the train. Frames of reference are particularly important when describing an object's displacement.

Frames of Reference is a 1960 educational film by Physical Sciences Study Committee. The film was made to be shown in high school physics courses. In the film University of Toronto physics professors Patterson Hume and Donald Ivey explain the distinction between inertial and non-intertial frames of reference, while demonstrating these concepts through humorous camera tricks. For example, the film opens with Dr. Hume, who appears to be upside down, accusing Dr. Ivey of being upside down. Only when the pair flip a coin does it become obvious that Dr. Ivey — and the camera — are indeed inverted. The film's humor serves both to hold students' interest and to demonstrate the concepts being discussed. This PSSC film utilizes a fascinating set consisting of a rotating table and furniture occupying surprisingly unpredictable spots within the viewing area. The fine cinematography by Abraham Morochnik, and funny narration by University of Toronto professors Donald Ivey and Patterson Hume is a wonderful example of the fun a creative team of filmmakers can have with a subject that other, less imaginative types might find pedestrian.

Displacement

Displacement is the change in position of an object relative to its reference frame. For example, if a car moves from a house to a grocery store, its displacement is the relative distance of the grocery store to the reference frame, or the house. The word "displacement" implies that an object has moved or has been displaced. Displacement is the change in position of an object and can be represented mathematically as follows:

$$\Delta x = x_f - x_0$$

where Δx is displacement, x_f is the final position, and x_0 is the initial position.

Shows the importance of using a frame of reference when describing the displacement of a passenger on an airplane.

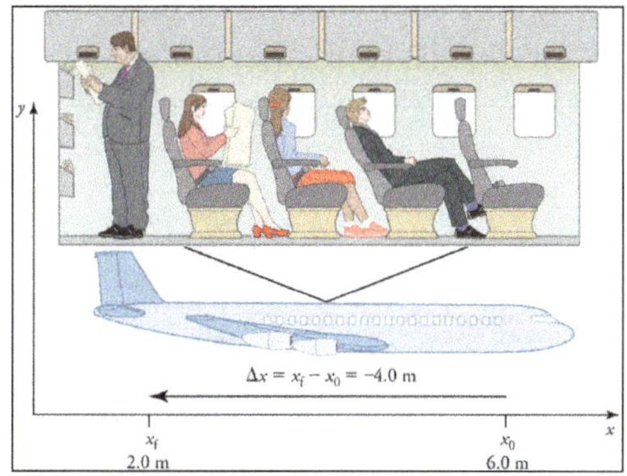

Displacement in Terms of Frame of Reference.

A passenger moves from his seat to the back of the plane. His location relative to the airplane is given by x. The -4.0m displacement of the passenger relative to the plane is represented by an arrow toward the rear of the plane. Notice that the arrow representing his displacement is twice as long as the arrow representing the displacement of the professor (he moves twice as far).

Scalars and Vectors

A vector is any quantity that has both magnitude and direction, whereas a scalar has only magnitude.

What is the difference between distance and displacement? Whereas displacement is defined by both direction and magnitude, distance is defined by magnitude alone. Displacement is an example of a vector quantity. Distance is an example of a scalar quantity. A vector is any quantity with both magnitude and direction. Other examples of vectors include a velocity of 90 km/h east and a force of 500 newtons straight down.

Scalars and Vectors: Mr. Andersen explains the differences between scalar and vectors quantities. He also uses a demonstration to show the importance of vectors and vector addition.

In mathematics, physics, and engineering, a vector is a geometric object that has a magnitude (or length) and direction and can be added to other vectors according to vector algebra. The direction of a vector in one-dimensional motion is given simply by a plus (+) or minus (–) sign. A vector is frequently represented by a line segment with a definite direction, or graphically as an arrow, connecting an initial point A with a terminal point B, as shown in figure.

Vector representation: A vector is frequently represented by a line segment with a definite direction, or graphically as an arrow, connecting an initial point A with a terminal point B.

Some physical quantities, like distance, either have no direction or no specified direction. In physics, a scalar is a simple physical quantity that is not changed by coordinate system rotations or translations. It is any quantity that can be expressed by a single number and has a magnitude, but no direction. For example, a 20 °C temperature, the 250 kilocalories (250 Calories) of energy in a candy bar, a 90 km/h speed limit, a person's 1.8 m height, and a distance of 2.0 m are all scalars, or quantities with no specified direction. Note, however, that a scalar can be negative, such as a −20 °C temperature. In this case, the minus sign indicates a point on a scale rather than a direction. Scalars are never represented by arrows. (A comparison of scalars vs. vectors is shown in figure.)

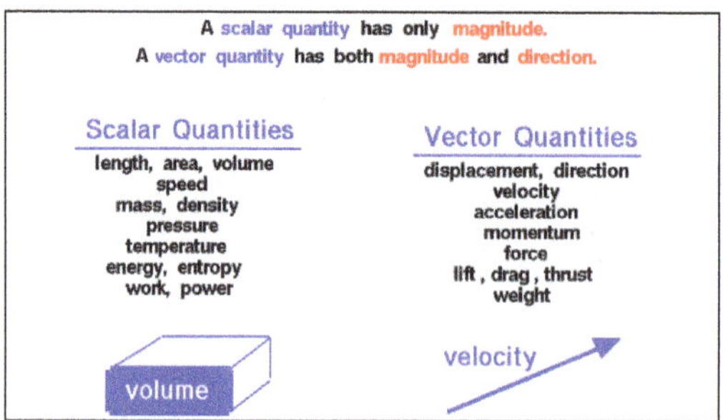

Scalars vs. Vectors: A brief list of quantities that are either scalars or vectors.

THERMODYNAMICS

Thermodynamics is the branch of physics that deals with the relationships between heat and other forms of energy. In particular, it describes how thermal energy is converted to and from other forms of energy and how it affects matter.

Thermal energy is the energy a substance or system has due to its temperature, i.e., the energy of moving or vibrating molecules, according to the Energy Education website of the Texas Education Agency. Thermodynamics involves measuring this energy, which can be "exceedingly complicated," according to David McKee, a professor of physics at Missouri Southern State University. "The systems that we study in thermodynamics consist of very large numbers of atoms or molecules interacting in complicated ways. But, if these systems meet the right criteria, which we call equilibrium, they can be described with a very small number of measurements or numbers. Often this is idealized as the mass of the system, the pressure of the system, and the volume of the system, or some other equivalent set of numbers. Three numbers describe 10^{26} or 10^{30} nominal independent variables."

Heat

Thermodynamics, then, is concerned with several properties of matter; foremost among these is heat. Heat is energy transferred between substances or systems due to a temperature difference between them, according to Energy Education. As a form of energy, heat is conserved, i.e., it cannot be created or destroyed. It can, however, be transferred from one place to another. Heat can also be converted to and from other forms of energy. For example, a steam turbine can convert heat to kinetic energy to run a generator that converts kinetic energy to electrical energy. A light bulb can convert this electrical energy to electromagnetic radiation (light), which, when absorbed by a surface, is converted back into heat.

Temperature

The amount of heat transferred by a substance depends on the speed and number of atoms or molecules in motion, according to Energy Education. The faster the atoms or molecules move, the higher the temperature, and the more atoms or molecules that are in motion, the greater the quantity of heat they transfer.

Temperature is "a measure of the average kinetic energy of the particles in a sample of matter, expressed in terms of units or degrees designated on a standard scale," according to the American Heritage Dictionary. The most commonly used temperature scale is Celsius, which is based on the freezing and boiling points of water, assigning respective values of 0 degrees C and 100 degrees C. The Fahrenheit scale is also based on the freezing and boiling points of water which have assigned values of 32 F and 212 F, respectively.

Scientists worldwide, however, use the Kelvin (K with no degree sign) scale, named after William Thomson, 1st Baron Kelvin, because it works in calculations. This scale uses the same increment as the Celsius scale, i.e., a temperature change of 1 C is equal to 1 K. However, the Kelvin scale starts at absolute zero, the temperature at which there is a total absence of heat energy and all molecular motion stops. A temperature of 0 K is equal to minus 459.67 F or minus 273.15 C.

Specific Heat

The amount of heat required to increase the temperature of a certain mass of a substance by a certain amount is called specific heat, or specific heat capacity, according to Wolfram Research. The conventional unit for this is calories per gram per kelvin. The calorie is defined as the amount of heat energy required to raise the temperature of 1 gram of water at 4 C by 1 degree.

The specific heat of a metal depends almost entirely on the number of atoms in the sample, not its mass. For instance, a kilogram of aluminum can absorb about seven times more heat than a kilogram of lead. However, lead atoms can absorb only about 8 percent more heat than an equal number of aluminum atoms. A given mass of water, however, can absorb nearly five times as much heat as an equal mass of aluminum. The specific heat of a gas is more complex and depends on whether it is measured at constant pressure or constant volume.

Thermal Conductivity

Thermal conductivity (k) is "the rate at which heat passes through a specified material, expressed as the amount of heat that flows per unit time through a unit area with a temperature gradient of one degree per unit distance," according to the Oxford Dictionary. The unit for k is watts (W) per meter (m) per kelvin (K). Values of k for metals such as copper and silver are relatively high at 401 and 428 W/m·K, respectively. This property makes these materials useful for automobile radiators and cooling fins for computer chips because they can carry away heat quickly and exchange it with the environment. The highest value of k for any natural substance is diamond at 2,200 W/m·K.

Other materials are useful because they are extremely poor conductors of heat; this property is referred to as thermal resistance, or R-value, which describes the rate at which heat is transmitted through the material. These materials, such as rock wool, goose down and Styrofoam, are used for insulation in exterior building walls, winter coats and thermal coffee mugs. R-value is given in units of square feet times degrees Fahrenheit times hours per British thermal unit ($ft^2 \cdot {}^\circ F \cdot h/Btu$) for a 1-inch-thick slab.

Newton's Law of Cooling

In 1701, Sir Isaac Newton first stated his Law of Cooling ("A Scale of the Degrees of Heat") in the Philosophical Transactions of the Royal Society. Newton's statement of the law translates from the original Latin as, "the excess of the degrees of the heat were in geometrical progression when the times are in an arithmetical progression." Worcester Polytechnic Institute gives a more modern version of the law as "the rate of change of temperature is proportional to the difference between the temperature of the object and that of the surrounding environment."

This results in an exponential decay in the temperature difference. For example, if a warm object is placed in a cold bath, within a certain length of time, the difference in their temperatures will decrease by half. Then in that same length of time, the remaining difference will again decrease by half. This repeated halving of the temperature difference will continue at equal time intervals until it becomes too small to measure.

Heat Transfer

Heat can be transferred from one body to another or between a body and the environment by three different means: conduction, convection and radiation. Conduction is the transfer of energy through a solid material. Conduction between bodies occurs when they are in direct contact, and molecules transfer their energy across the interface.

Convection is the transfer of heat to or from a fluid medium. Molecules in a gas or liquid in contact with a solid body transmit or absorb heat to or from that body and then move away, allowing other molecules to move into place and repeat the process. Efficiency can be improved by increasing the surface area to be heated or cooled, as with a radiator, and by forcing the fluid to move over the surface, as with a fan.

Radiation is the emission of electromagnetic (EM) energy, particularly infrared photons that carry heat energy. All matter emits and absorbs some EM radiation, the net amount of which determines whether this causes a loss or gain in heat.

The Carnot Cycle

In 1824, Nicolas Léonard Sadi Carnot proposed a model for a heat engine based on what has come to be known as the Carnot cycle. The cycle exploits the relationships among pressure, volume and temperature of gasses and how an input of energy can change form and do work outside the system.

Compressing a gas increases its temperature so it becomes hotter than its environment. Heat can then be removed from the hot gas using a heat exchanger. Then, allowing it to expand causes it to cool. This is the basic principle behind heat pumps used for heating, air conditioning and refrigeration.

Conversely, heating a gas increases its pressure, causing it to expand. The expansive pressure can then be used to drive a piston, thus converting heat energy into kinetic energy. This is the basic principle behind heat engines.

Entropy

All thermodynamic systems generate waste heat. This waste results in an increase in entropy, which for a closed system is "a quantitative measure of the amount of thermal energy not available to do work," according to the American Heritage Dictionary. Entropy in any closed system always increases; it never decreases. Additionally, moving parts produce waste heat due to friction, and radiative heat inevitably leaks from the system.

This makes so-called perpetual motion machines impossible. Siabal Mitra, a professor of physics at Missouri State University, explains, "You cannot build an engine that is 100 percent efficient, which means you cannot build a perpetual motion machine. However, there are a lot of folks out there who still don't believe it, and there are people who are still trying to build perpetual motion machines."

Entropy is also defined as "a measure of the disorder or randomness in a closed system," which also inexorably increases. You can mix hot and cold water, but because a large cup of warm water is more disordered than two smaller cups containing hot and cold water, you can never separate it back into hot and cold without adding energy to the system. Put another way, you can't unscramble an egg or remove cream from your coffee. While some processes appear to be completely reversible, in practice, none actually are. Entropy, therefore, provides us with an arrow of time: forward is the direction of increasing entropy.

The Four Laws of Thermodynamics

The fundamental principles of thermodynamics were originally expressed in three laws. Later, it was determined that a more fundamental law had been neglected, apparently because it had seemed so obvious that it did not need to be stated explicitly. To form a complete set of rules, scientists decided this most fundamental law needed to be included. The problem, though, was that the first three laws had already been established and were well known by their assigned numbers. When faced with the prospect of renumbering the existing laws, which would cause considerable confusion, or placing the pre-eminent law at the end of the list, which would make no logical sense, a British physicist, Ralph H. Fowler, came up with an alternative that solved the dilemma: he called the new law the "Zeroth Law." In brief, these laws are:

- The Zeroth Law states that if two bodies are in thermal equilibrium with some third body, then they are also in equilibrium with each other. This establishes temperature as a fundamental and measurable property of matter.

- The First Law states that the total increase in the energy of a system is equal to the increase in thermal energy plus the work done on the system. This states that heat is a form of energy and is therefore subject to the principle of conservation.

- The Second Law states that heat energy cannot be transferred from a body at a lower temperature to a body at a higher temperature without the addition of energy. This is why it costs money to run an air conditioner.

- The Third Law states that the entropy of a pure crystal at absolute zero is zero. As explained above, entropy is sometimes called "waste energy," i.e., energy that is unable to do work, and since there is no heat energy whatsoever at absolute zero, there can be no waste energy. Entropy is also a measure of the disorder in a system, and while a perfect crystal is by definition perfectly ordered, any positive value of temperature means there is motion within the crystal, which causes disorder. For these reasons, there can be no physical system with lower entropy, so entropy always has a positive value.

The science of thermodynamics has been developed over centuries, and its principles apply to nearly every device ever invented. Its importance in modern technology cannot be overstated.

STRUCTURAL ANALYSIS

Structural analysis is a comprehensive assessment to ensure that the deformations in a structure will be adequately lower than the permissible limits, and failure of structural will not occur. The aim of structural analysis is to design a structure that has the proper strength, rigidity, and safety. Deformations in a structure can be either elastic that is totally recoverable, or inelastic that is permanent. Structural analysis assists in the design of structures that meet their functional requirements, are economical and attractive. Structural analysis integrates the disciplines of mechanics, dynamics, and failure theories to compute the internal forces and stresses on the structures to be designed.

Modes of Structural Analysis

Structural analysis is carried out by an examination of the real structure, on a model of the structure created on some scale, and by the utilization of mathematical models. Tests are conducted on the real structure when production is required of similar structures in large quantities, like frames of a particular car, or when the test expenses are acceptable due to the significance of the task. When elements of the main structures are to be examined, then models are used for the estimation of the different loads to be endured. Most structural analyses are conducted on the mathematical models, in which the model could be elastic or inelastic, forces may be static or dynamic, and the model of the structure might be two dimensional or three dimensional.

Strength of Materials

There are several methods to determine the stresses in different structural members, such as columns and beams that may cause failure under loading. To conduct an accurate structural analysis, complete data concerning the structural loads, properties of materials, and stresses involved is essential. The approach related to the strength of materials is utilized for simple structural elements subjected to plain loads, such as shafts subjected to torsion, and axial loading of bars. The strength of a material is dependent upon its microstructure that can be changed by the application of various engineering techniques on the material, such as work hardening and grain boundary hardening. However, various mechanical characteristics of a material may deteriorate in an effort to increase the strength, such as when the grain size is decreased, the yield strength is increased but the material becomes brittle.

Estimation of Loads

One important element of the structural analysis is the accurate analysis of the estimation of structural loads to be endured. Structural loads are the forces on a part of a structure, or on the complete structure, and evaluation of their consequences is conducted by structural analysis. Overload loading may result into failure of structures, and such conditions should be considered during the design of the structures. For buildings and bridges, the main vertical loads are gravity loads, including the structure weight and the permanent accessories, known as dead loads. The live loads are the concentrated loads, or distributed loads over large areas such as floors. The horizontal loading on buildings is due to wind and inertial forces caused by earthquakes.

References

- Mechanics, science: britannica.com, Retrieved 26 June, 2019
- Basics-of-kinematics, chapter, boundless-physics: lumenlearning.com, Retrieved 30 August, 2019
- Thermodynamics-50776: livescience.com, Retrieved 15 April, 2019
- Structural-engineering, basics-of-structural-analysis-44070: brighthubengineering.com, Retrieved 17 May, 2019

4

Mechanical Vibrations

Mechanical vibration refers to the measurement of a periodic process of oscillations in relation to an equilibrium point. Some of the important areas of study within this field are vibration theory, string vibration and forced vibrations. All these diverse concepts of mechanical vibrations have been carefully analyzed in this chapter.

Any oscillation motion of mechanical system about its equilibrium position is called vibration. The simplest example of vibration system is mass-spring system. Body represents by a mass starts vibrate when it is displaced from a position of stable equilibrium. Body keep moving back and forth across its equilibrium caused by restoring force, in this case restoring forces is elastic force of spring. This elementary vibration system is called mechanical oscillator.

Mass-spring system performs harmonic periodic motion with constant amplitude of displacement if the system is undamped. Undamped system conserves mechanical energy – harmonic motion will carry on vibrating forever. There will be continuous interchange of kinetic to potential energy and back again. But undamped system is simplifying of reality because in the real world is impossible to isolate the system – friction and air-resistance drains mechanical energy from system into environment in form of heat energy. Total mechanical energy left in the system therefore gradually decreases – amplitude of displacement also decreases in time.

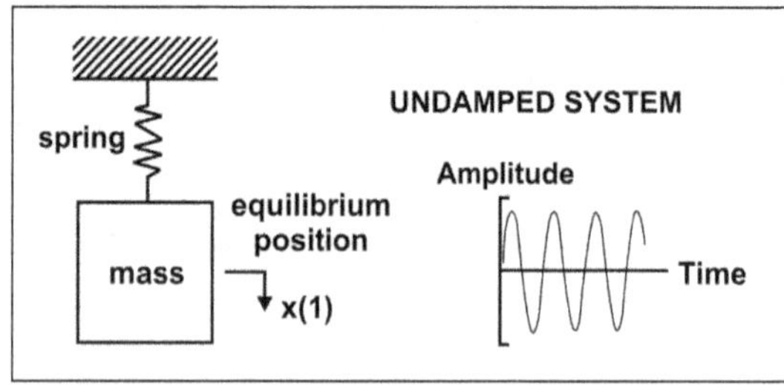

Undamped mechanical oscillator.

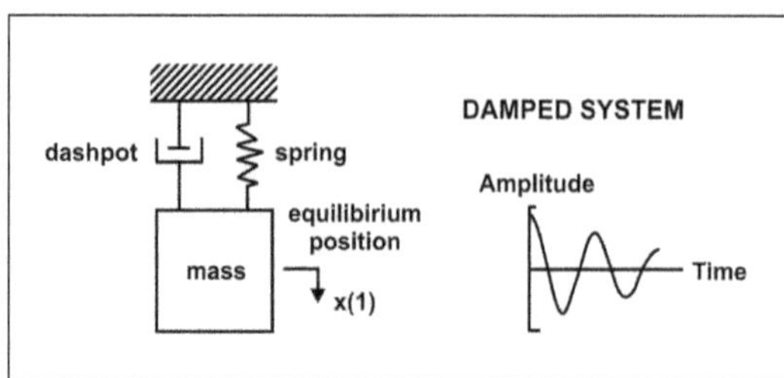

Damped mechanical oscillator.

Other elementary mechanical oscillators are shown in figure:

- Simple pendulum: restoring force is gravitational force.
- Bending beam: restoring force is elasticity of beam material.
- Torsion of shaft: restoring force is elasticity of shaft material.

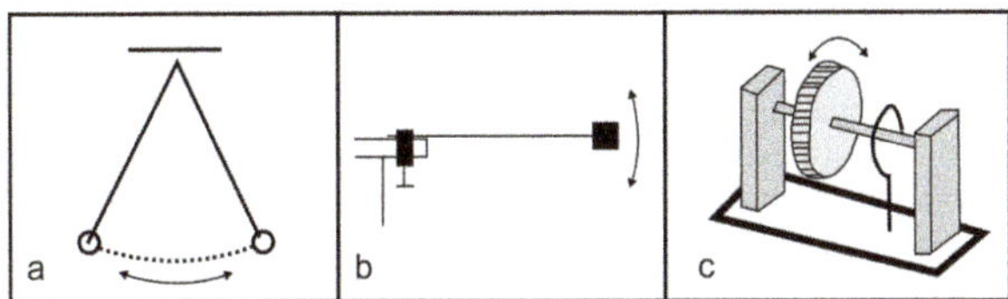

Mechanical oscillators.

Almost every machines produce vibrations, for example combustion engines vibrate caused by periodic movement of pistons, rotational devices vibrate caused unbalanced parts; automobiles vibrate caused by road surface roughness, etc. Human speech is a product of vocal cord vibration, recognize sounds is caused by oscillation of the eardrum, operation of many musical instruments is based on vibration.

Alternative current (AC) is the result of periodic oscillations of electric charges. If AC flows through induction coil the magnetic field in coil's cavity is changed. AC also changes electric field in electrical capacitor. These examples show that vibration is not phenomenon only in mechanical engineering but it is an element in many other physical domains (electricity, magnetism, etc).

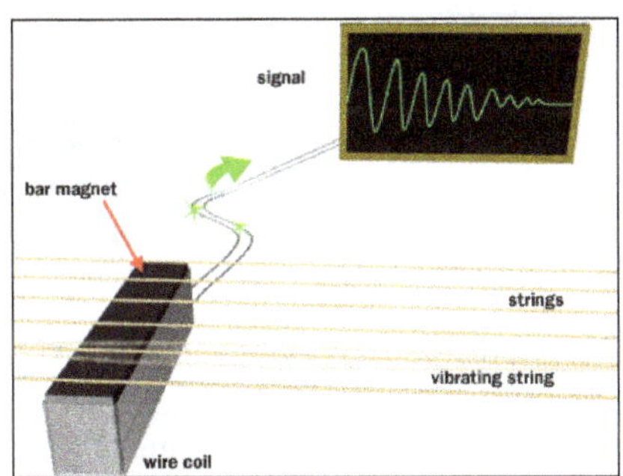

Guitar string vibration.

Vibrations can be classified into three categories:

- Free vibration of a system is vibration that occurs in the absence of external force. The sources of free vibration are initial displacement of system from equilibrium or give the initial velocity to the system.

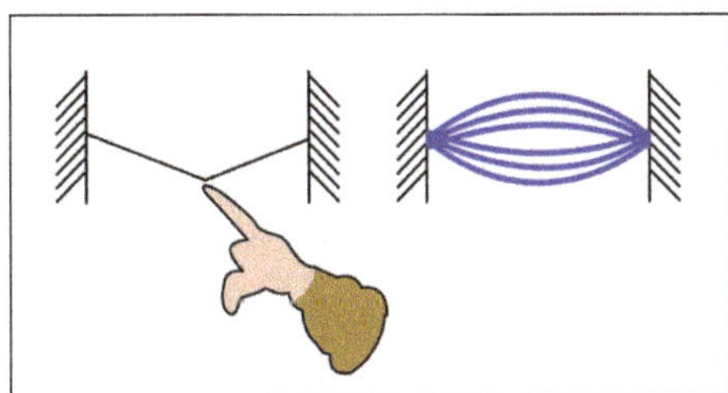

Free vibration: String vibration.

- Forced vibration is caused by an external force that acts on the system. In this case, the exciting force continuously supplies energy to the system.

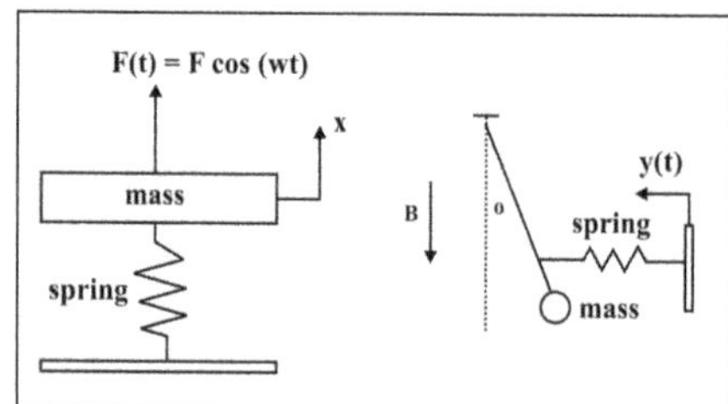

Forced vibration: system forced by a harmonic external force (left), system forced by a time-dependent displacement (right).

- Self-excited vibration is periodic and deterministic oscillation. Under certain conditions, the equilibrium state in such a vibration system becomes unstable, and any disturbance causes the perturbations to grow until some effect limits any further growth. In contrast to forced vibrations, the exciting force is independent of the vibrations and can still persist even when the system is prevented from vibrating. The force acting on a vibrating object is usually external to the system and independent of the motion. However, there are systems in which the exciting force is a function of the motion variables (displacement, velocity or acceleration) and thus varies with the motion it produces. Friction-induced vibration (in vehicle clutches and brakes, vehicle-bridge interaction) and flow-induced vibration (circular wood saws, CDs, DVDs, in machining, fluid-conveying pipelines), fluttering of airplane wing are examples of self-excited vibration.

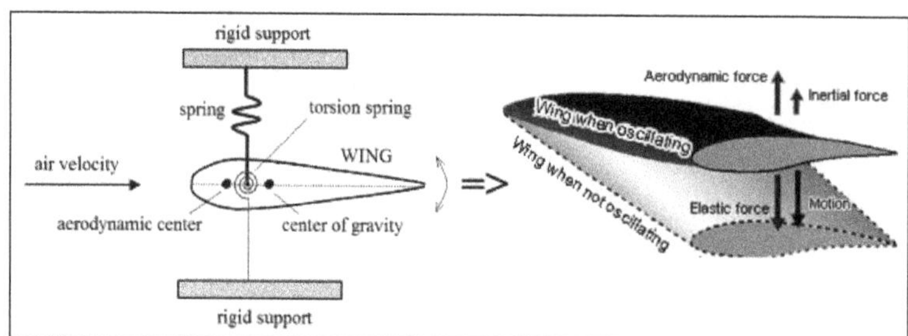

Self-excited vibrations: fluttering of wing.

Most vibrations are undesirable in mechanical engineering. Vibrations in machines and structures because produce increased stresses, energy losses, cause added wear, increase bearing loads, induce fatigue, create passenger discomfort in vehicles, and absorb energy from the system . Rotating machine parts need careful balancing in order to prevent damage from vibrations . The worst impact has resonance of mechanical systems. Resonance can occur when forced vibration and can cause even at low loads serious damages. Understanding of vibrations is therefore very important for engineers.

Mechanical systems is general consist of structural components which have distributed mass and elasticity. Examples of these structural components are rods, beams, plates, and shells. These structural components are considered as continuum systems which have an infinite number of degrees of freedom (DOF) and therefore the vibration of real systems is governed by partial differential equations which involve variables that depend on time as well as the spatial coordinates. For study of vibration is preferred by simplifying real system to discrete system with a finite number of DOF. Discrete system is represented by lumped mass and discrete elastic elements (translational and torsion springs) and discrete damped elements (viscous dashpots). Physical models of discrete systems are shown in figure. These systems are governed by a set of second-order ordinary differential equations.

The cantilever beam is an example of a continuum system. Under certain conditions this beam can be modeled as a simple discrete spring-mass system. In order to model the vibration of the cantilever beam, the end of the beam is chosen as a reference point at which the characteristics and response of the beam are measured. An equivalent system is then built that has a response – y(t), that is identical to that of the actual system.

The spring constant – k, of the equivalent system is identical to that of the cantilever beam, and can be calculated quite easily using beam deflection formulae. Calculation of an equivalent mass is necessary because all points along the beam's length do not have the same response as the end of the beam. This means that the equivalent mass – m, cannot be determined simply by adding the masses m_{beam} and m_{end}, but must be found by equating the energies of the two systems as they vibrate (this type of analysis is called lumping).

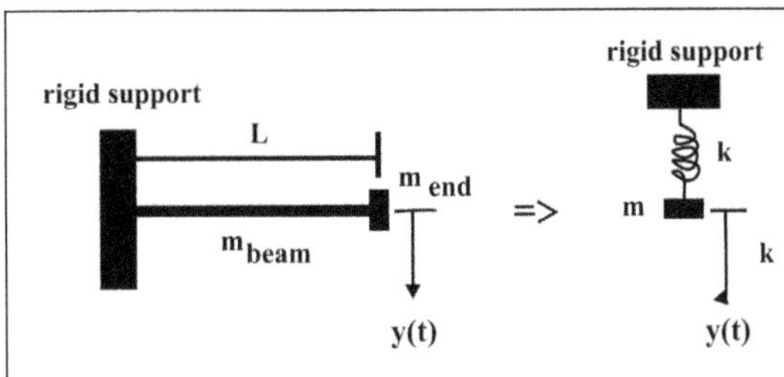

Antilever beam as a simple spring-mass system.

VIBRATION THEORY

Vibrations are a common occurrence in oil and gas activities that can affect operations, planning, facility design, and interpretation of results. Vibration is common in drillstrings, on platforms, wherever large engines are operating, in seismic operations, and many other aspects of oil and gas. Understanding vibration theory and the mathematics of vibrations are important to successful operations. A refresher on differential calculus can come in handy as well.

The fundamental theories of vibration are not new. Indeed, Saint-Venant published his theory on the vibrations of rods in 1867, and Love published an entire treatise on vibration theory in 1926. The mathematics of vibration theory involves infinite series, complex functions, and Fourier integral transforms, and its physics involves Newtonian mechanics and stress analyses. Until recently, except under relatively simple conditions, the complexity of such mathematics had restrained the application of vibration theory to solving simple common problems. Now, however, state-of-the-art computers can perform these complex calculations in a reasonable time frame, making possible a wave of new studies.

A vibration is a fluctuating motion about an equilibrium state. There are two types of vibration: deterministic and random. A deterministic vibration is one that can be characterized precisely, whereas a random vibration only can be analyzed statistically. The vibration generated by a pumping unit is an example of a deterministic vibration, and an intermittent sticking problem within the same system is a random vibration.

In mechanical systems, deterministic vibrations are excitations that elicit a response from a system, as shown schematically in figure. In theory, as long as two of the three variables (excitation, system, and response) are known, the third one can be determined; however, the mathematics might be challenging. Most often, the response function is sought, so that the excitation function and the system must be known.

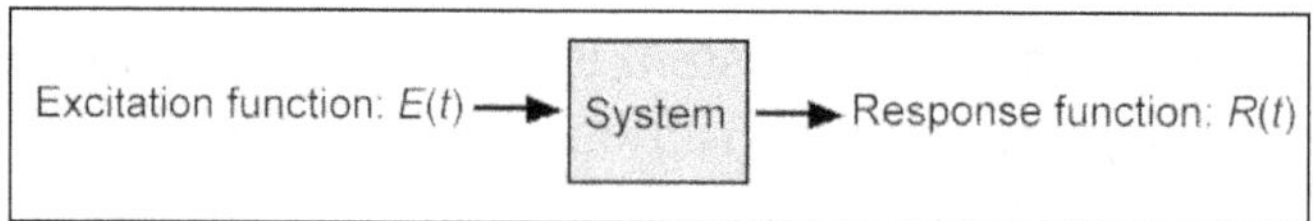

Excitation/response System for Deterministic Vibrations.

Vibration systems can be linear or nonlinear, and discrete or continuous. In all cases, a vibration system can be in one, two, or three mutually orthogonal dimensions. A linear system is a system in which proportionality $\text{if } E(t) \to R(t), \text{then } r\,E(t) \to rR(t)$ and superposition $E_1(t) \to R_1(t), \text{and } E_2(t) \to R_2(t)$ are true, that is, in which:

$$\text{if } E(t) \to R(t), \text{then } r\,E(t) \to rR(t)$$

and

$$\text{if } E_1(t) \to R_1(t), \text{and } E_2(t) \to R_2(t),$$

When proportionality and superposition are not true, then the system is nonlinear.

	Linear	Nonlinear
Discrete	One DOF Two DOFs Three DOFs	One DOF Two DOFs Three DOFs
Continuous	1D, 2D, or 3D	1D, 2D, or 3D

Vibration System Classification.

A discrete system is one having a finite number of independent coordinates that can describe a system response. These independent coordinates are known as degrees of freedom (DOFs). If the motion of mass, either translational or rotational, of a vibrating system is a function of only one independent coordinate, then the system has one DOF. If two or more independent coordinates are required to describe one or both types of motion, then the system has two or more DOFs. If a system is continuous (an infinite set of independent coordinates is needed to describe the system response), it has an infinite number of DOFs. Because material structures all have a continuous nature, all systems have an infinite number of DOFs. Most systems have dominant DOFs; some even have a single dominant DOF. Such systems therefore can be characterized as discrete systems, which makes the mathematics more tractable.

If a system has a single DOF or set of DOFs in only one direction, it is a 1D system. If there are two mutually orthogonal directions for the DOF, it is a 2D system; and if there are three mutually orthogonal directions for the DOF, it is a 3D system.

As figure shows, the excitation function can be periodic or transient, and absent or present. A periodic vibration is one that can be characterized mathematically as an indefinite repetition. A transient vibration is of finite length and is composed of waves that have a definite beginning and that eventually die out. These waves can be of extremely short duration or last for some time.

	Linear	Nonlinear
Discrete	One DOF Two DOFs Three DOFs	One DOF Two DOFs Three DOFs
Continuous	1D, 2D, or 3D	1D, 2D, or 3D

Excitation Function Classification.

A standing wave is a vibration whose wave profile appears to be standing still, though actually the particles that make up the material are oscillating about an equilibrium position. Because of the geometry and boundary conditions of the material through which they are traveling, the waves and the reflected waves cancel and reinforce themselves over the same location in the material, which makes the wave profile appear not to be moving. The point at which no motion is occurring is a nodal point, or node. The point of maximum amplitude is the antinode.

In reality, all waves are transient in some way. If a wave is repeated over a longer time than it takes for a single wave to propagate through a material, then this series of waves can be called a vibration. All vibrations are transient, as well. If the vibration lasts longer than the time under analysis, then it can be characterized as infinite in length.

When the excitation is present and is actively affecting the system within the analysis time frame, the response is called a forced vibration. The response of a system with an absent excitation function—one that is not present within the analysis time frame—is called a free vibration. As such, the system can be responding to the removal of an excitation function. For example, if the response of a mass and spring system is sought after the system has been pulled down and released, the original excitation function (the pulling force) is considered absent because the analysis is being performed after the release.

Wave Propagation

The method by which a vibration travels through a system is known as wave propagation. When an external force is impressed on a real-world elastic body, the body does not react instantly over its entire length. The point immediately under the external force reacts first, and then the section just under that point reacts to the previous section's reaction, and so on. This series of reactions is called wave propagation because the reactions propagate through the body over a period of time at a specific velocity. If the rate of change of the external force is slow enough, static equilibrium analysis can model the reactions adequately for most engineering applications. This is called rigid-body analysis. If the external force changes rapidly, however, wave-propagation analysis is necessary to model the reactions effectively.

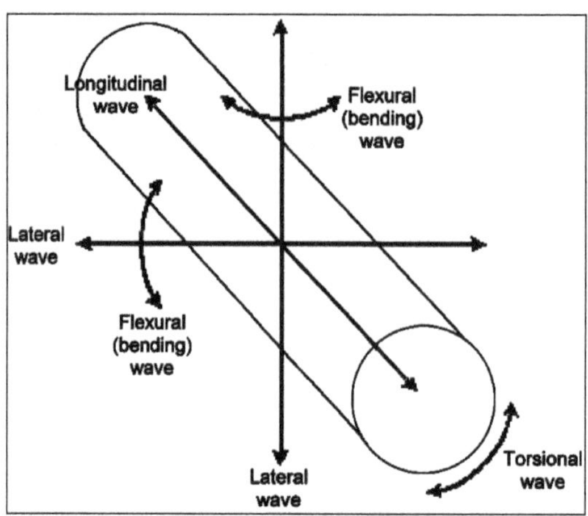

Types of Elastic Waves.

There are many types of elastic waves. In longitudinal waves (also variously called compression/tension, axial, dilatational, and irrotational waves), the particles that make up the elastic medium are forced directly toward and away from each other, and the direction of the particles' motion is parallel to that of the wave motion. In most steels, longitudinal waves travel at ≈16,800 ft/sec. Longitudinal waves are not dispersive. This means that all the wave components that make up a longitudinal wave travel at the same velocity and, hence, do not separate (disperse).

In lateral waves (also known variously as shear, torsional, transverse, equivoluminal, and distortional waves), the particles slip beside each other, and move perpendicular to the direction of the wave motion. Because slipping uses more energy, lateral waves are slower. In steel, for example, they travel at 10,400 ft/sec. A rapidly changing torsional force on a section of pipe will cause a lateral wave to propagate from the point of application to all other parts of the pipe. It propagates as an angular twist. Lateral waves are nondispersive and have a similar solution method as the longitudinal waves; however, shear or transverse waves are dispersive (i.e., the wave components that make up a shear wave travel at different velocities). Their wave components will disperse and "smear" the initial wave profile. This complicates the analysis significantly.

Bending waves (flexural waves) travel as a bend in a bar or plate and have longitudinal and lateral components. Rotary shears and moments of inertia complicate bending-wave analysis.

Wave-propagation studies in petroleum engineering areas generally have been confined to longitudinal, torsional, and lateral waves. Of these, longitudinal waves generally are easiest to model. A compression wave is a stress wave in which the propagated stress is in compression. Likewise, a tension wave is a stress wave in which the propagated stress is in tension.

Wave Reflection from Various Geometric Boundaries

Wave propagation is the movement of a distinct group of waves through some material in response to an external force.

A key point in wave-propagation studies is how waves interact with geometric discontinuities. What happens as a wave meets a fixed or free boundary condition? Also, what happens to a wave as it encounters a geometrical area change or a change in material properties?

There are two limiting boundary conditions for wave propagation: a fixed (pinned) end (zero displacement) and a free end (zero stress). A fixed end is a boundary condition in which there is zero displacement. According to wave theory, during a wave encounter with a fixed end, the stress at the fixed end doubles during the passage of the wave. A reflection of a stress wave will simply bounce back with the same sign. A compression wave will reflect as a compression wave and a tension wave will reflect as a tension wave. At a fixed end, because the displacement is zero, the particle velocity will be zero. The wave particle velocity amplitude is inverted during a reflection from a fixed end.

A free end is defined as a boundary condition free to move. The stress at the free end is always zero. The effects on stress and particle velocity caused by a free end are opposite of the effects on stress on a fixed end. A compression wave encountering a free end reflects as a tension wave, and a tension wave reflects as a compression wave. The wave particle velocity values double during an encounter with a free end and reflect with the same sign.

As a wave encounters a change in cross-sectional area, some of the wave is reflected and some is transmitted (refracted). The amplitudes and sign of the waves depend on the relative change in cross-sectional area. The equation that describes the effect on the incident force, F_i, of a cross-sectional-area, density, or modulus-of-elasticity change for the transmitted wave is,

$$F_t = \frac{\sqrt{\dfrac{E_2\rho_2}{E_1\rho_1}}}{\sqrt{\dfrac{E_2\rho_2}{E_1\rho_1}\dfrac{A_{c2}}{A_{c1}}}+1} F_i,$$

and for the reflected wave is,

$$F_r = \frac{\sqrt{\dfrac{E_2\rho_2}{E_1\rho_1}\dfrac{A_{C2}}{A_{c1}}}-1}{\sqrt{\dfrac{E_2\rho_2}{E_1\rho_1}\dfrac{A_{C2}}{A_{c1}}}+1} F$$

If an incident wave encounters a junction where the relative change in cross-sectional area is greater than 1 (a smaller area to a larger area), most of the wave will transmit through the junction. Some of the wave will reflect from the junction and will keep the same sign. For example, a compression wave will transmit through the junction and keep going as a somewhat-diminished compression wave. The part of the wave that is reflected is still a compression wave, but its amplitude is less than that of the wave that transmitted though the junction.

On the other hand, if an incident wave encounters a junction where the relative change in cross-sectional area is less than 1 (a larger area to a smaller area), most of the wave will reflect off the junction, but some of it will transmit through the junction and will keep the same sign. For example, a compression wave will transmit through the junction and keep going as a diminished compression wave. The reflected part of the wave is a tension wave whose absolute amplitude is greater than that of the compression wave that is transmitted through the junction.

As with most drillstrings, there are many geometric discontinuities (changes in cross-sectional area) that will cause part of the wave to refract and part to reflect. For example, drill collars to heavyweight drillpipe to drillpipe all are geometric discontinuities. Sometimes, too, there are material discontinuities—changes in material density or modulus of elasticity—that cause refractions and reflections. A third possible type of discontinuity is when there are different endpoints. For example, if the pipe is stuck, one end can be modeled as stuck. If the pipe is hanging freely, such as with casing running, then the end is free.

Wave Behavior

Wave velocity depends primarily on density and modulus of elasticity but also is affected by damping and frequency. For example, hitting one end of a long steel rod with a hammer will generate a longitudinal wave that compresses the particles of the steel. The wave's length is set by the length of time that the hammer is in contact with the end of the rod, whereas its magnitude is set by the force of the hammer blow. As the wave moves along the rod, the steel within the length of the wave is compressed. After the wave passes, the steel returns to its unstressed state, though not necessarily in the same location as before the wave passage.

As another example, twisting (shearing) a steel rod will generate a shear wave. A shear wave moves along the rod more slowly than the longitudinal wave does. Similarly to the longitudinal wave discussed above, its length is set by the duration of the twisting action, whereas its magnitude is set by the torque from the twisting action.

Waves act independently, but the stresses they create can be additive. For example, two equal compression waves that are generated simultaneously by hammer blows at each end of a long steel rod will meet in the center of the rod, pass through one another, and then each continue along the rod as if the other never existed (independence). While the waves are passing each other, however, the compression in the steel will be twice (additive) that of either wave.

Natural Frequencies and Resonance

Everything has a natural frequency, a frequency at which it would vibrate were it given the energy to vibrate and left alone. For instance, the human body has a natural frequency of ≈5 cycles/sec. All drill and rod strings have a natural frequency that depends on the material properties and geometry. The material properties determine the wave velocity, and the geometry determines how waves are reflected and refracted.

During wave propagation, the wave eventually reaches an end of the material. Some of the wave will reflect back to its source. If the reflection reaches the source at the same time a new wave is generated, the two waves will combine and be synchronized in phase. Later, if those two waves' reflections return to the source at the same time the next new wave is generated, all three waves will combine. This will continue for as long as waves are generated under these conditions, and the resultant wave will increase in amplitude, theoretically to infinity. This is called resonance. The frequency at which resonance occurs is the natural frequency or an integer multiple of that frequency (called a harmonic). If this wave reinforcement is allowed to continue, the system eventually will either self-destruct or fatigue to failure.

A continuous system contains an infinite number of natural frequencies, whereas a discrete single-degree-of-freedom (SDOF) system (e.g., a point mass on a massless spring) has only one natural frequency. If two point masses are connected using two springs, then there are two natural frequencies in this 2DOF system. In general, the number of DOFs in a system determines the number of natural frequencies it has, which means that any discrete system will have a finite number of natural frequencies; however, in reality, there is an infinite number of natural frequencies because all systems are continuous. Some frequencies will have higher amplitudes than others. Such continuous systems with discrete higher-amplitude responses can be modeled with a discrete methodology.

Damping

Resonance energy does not reach an infinite value because of damping, the dissipation of energy over time or distance. Without damping, or friction, the energy from vibrations would build until there is more energy than the structure can sustain, which can cause structural failure.

A wave propagating into a system adds energy to a system, whereas damping removes it. Generally, the dissipated energy from the vibration is converted to heat, and if damping does not take enough energy out of a system, the system can self-destruct from energy overload. The amount of energy in a system at a given time is reflected in the system's stress/strain level. The more stresses/strains in the system, the higher the energy level. Once the stresses reach a value greater than the yield strength of the system, yield failure is imminent. If the stresses are greater than the ultimate strength of the material, failure is immediate.

In the borehole, three distinctive types of damping occur: viscous, coulomb, and hysteretic. Viscous damping occurs when the damping force generated is proportional to the velocity of the particles. Coulomb damping (also called dry friction) is the force generated by the movement of materials past one another, and it usually is proportional to the force normal to the materials' surfaces. The dynamic and static coefficients of friction are the proportionality constants. Hysteretic damping is the friction force generated by the relative motion of the internal planes of a material as a wave causes particle motion. Although this is true of all materials, some materials are viscoelastic (i.e., they show a much larger hysteretic effect than do others).

Viscous Damping

As noted above, viscous damping occurs when the damping force is proportional to the velocity of the particles. Viscous damping is shown by:

$$F_d = c\frac{dx}{dt}$$

One way that viscous damping arises in jarring analysis is from the interaction of a solid and liquid at their interface, such as where the steel contacts the liquid mud along the sides of a drillstring.

One method for determining the damping involves noting the decrement of acceleration over one vibration cycle. An impulse is impressed on the drillstring to produce a wave. While the wave is decaying, the acceleration is measured and recorded multiple times at one location on the string and at the same phase (i.e., crest to crest). The time between recordings also is noted. These values are used in equation $c = 2\frac{A_c E}{V_s 2t}\ln\frac{a_1}{a_2}$ to compute the damping coefficient (c). Unfortunately, though, this method gives the total damping and does not distinguish between viscous and Coulomb damping.

$$c = 2\frac{A_c E}{V_s 2t}\ln\frac{a_1}{a_2}$$

Coulomb damping is the friction that occurs when two dry surfaces slide over each other, and its force is a constant value that is independent of particle velocity and displacement, but dependent on the friction factor (μ) and the force normal to the friction surface. This value is:

$$F_f = \mu F_n$$

The Coulomb damping force always is of the opposite sign from that of the particle velocity, so that the damping force reverses when the particle velocity changes signs. This discontinuity makes it a nonlinear damping force, shown as:

$$F_f = \pm\left(\frac{dx}{dt}\right)\mu F_n$$

Nonlinearity makes a closed-form solution to an equation of motion difficult.

Hysteretic Damping

Hysteretic damping also is called structural damping because it arises from internal friction within a structure. A wave moves through a material because the atomic structure is reacting to an applied force. As the atoms of the structure move, energy is lost through the interaction of these atoms with their neighboring atoms. Hysteretic damping is the energy lost when atoms move relative to each other.

If a material had a perfectly linear stress/strain relationship, hysteretic damping would not occur. In reality, though, there is no such thing as a perfectly linear stress/strain curve. Two curves develop on the stress/strain diagram while a material is stressed and relieved. The center area between these two curves represents the energy lost to internal friction. (This hysteresis loop is the reason for the name of this damping type.) This variation can be small, but the amount of energy dissipated can be large because high-frequency vibrations can cause this loop to be repeated many, many times over a given time period.

The hysteretic-damping value is highly dependent on a number of factors. One factor is the condition of the material (i.e., chemical composition, inhomogeneities, and property changes caused by thermal and stress histories). Another is the state of internal stress from initial and subsequent thermal and stress histories. Also, the type and variation of stress—axial, torsional, shear, and/or bending—affect the hysteretic-damping value.

A way of looking at hysteretic-damping force is to set it proportional to the particle velocity divided by the wave frequency. This is shown in equation below:

$$F_h = \frac{h}{\omega}\frac{dx}{dt}$$

Equivalent Springs

Many systems can be modeled as multiple springs. Such springs can be combined into a single, equivalent spring. For parallel springs, the sum of the spring constants is equal to the equivalent spring constant. For series springs, the reciprocal of the sum of the reciprocals of the spring constants is equal to the equivalent spring constant. A linear spring oscillates in a single translational direction. A torsional spring oscillates with an angular twist.

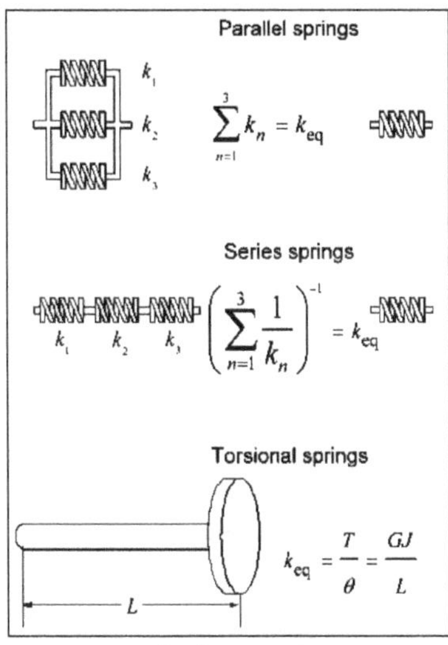

Equivalent Springs.

Boundary and Initial Conditions

The boundary conditions (how the ends of a system are attached) and initial condition (condition of the system at the start in time) are extremely important in vibration and wave propagation analysis. The specific solution of any ODE or PDE requires a set of boundary and/or initial conditions. Usually, a displacement (boundary condition) and an initial velocity (initial condition) are specified.

In wave propagation, the boundary conditions also dictate wave behavior. For example, a compression wave is reflected from a free end as a tension wave and from a fixed end as a compression wave. If two rods are connected at their ends and are of different geometry or material, then a fraction of the energy of the wave is reflected and the remaining portion of the energy is refracted at their connection. Other types of boundaries direct the system response by limiting the DOF. This includes:

- Boundary conditions of pinned, revolute, translational, translational and rotational.
- Forcing function.
- Mass spring and/or damper, and a semi-infinite connection.

In addition, changes in material properties will affect the various constants and will cause wave-propagation reflections and refractions at the boundary between the properties. The figure below shows some typical boundary conditions.

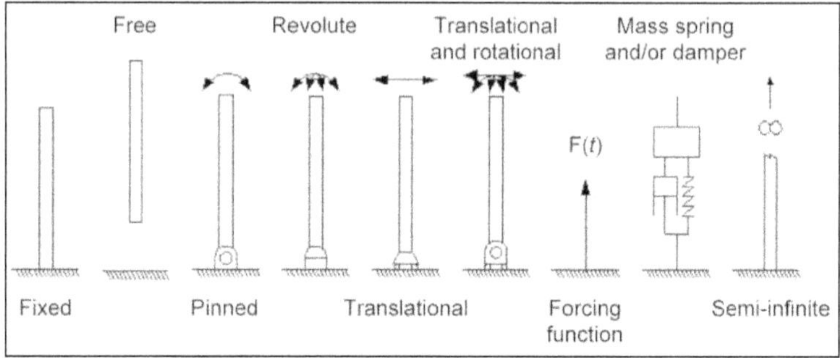

Types of Boundary conditions.

Mechanical Vibration Analysis

There are three components to mechanical vibration analysis:

- Determine the geometric compatibilities.
- Determine the constitutive (material properties) equations.
- Determine the equilibrium condition.

The geometric compatibilities are the displacement constraints and connections. They also include the continuous properties, which state that the system does not separate into individual pieces. (If it does, that is another problem altogether.) The constitutive equations represent the material properties, which include mass, damping, and spring coefficients. These constitutive equations include stress/strain relationships and Hooke's law:

$$\sigma = E\varepsilon$$

or, in another form,

$$F = \frac{EA_c}{L}\Delta l$$

The coefficient of Δl in equation $\sigma = E\varepsilon$ often is called the spring constant or stiffness constant.

The equilibrium condition is based on both static and dynamic conditions. A static equilibrium states that the sum of the forces acting on an object is equal to zero:

$$\sum F = 0.$$

A dynamic equilibrium is based on Newton's second law and is the basis of many vibration analysis methods. The sum of the forces acting on an object is equal to its mass times the acceleration of the object. Other dynamic-equilibrium analysis includes virtual work methods and energy-balance methods (Hamilton's principle).

Newton's second law for a translational system is,

$$\sum F = m\ddot{x}$$

and for torsional systems is,

$$\sum T = 1\ddot{\theta}$$

Newton's second law can be rewritten in a form known as D'Alembert's principle:

$$\sum F - m\ddot{x} = 0$$

in which $m\ddot{X}$ is treated as a force and is called an inertial force.

Some basic equations of vibration analysis are shown in table.

Item	Mathematical definition
Period	$\tau = 1/f = 2\pi/\omega$
Wavelength	$\lambda_w = 2\pi v_w/\omega = 2\pi/k_w$
Cyclic frequency	$f = \omega/2\pi$
Angular frequency	$\omega = 2\pi f$
Wave phase	$\phi = k_w x - \omega t = \omega/vw\,(x - v_w t) = 2\pi/\lambda w\,(X - v_w t)$
Phase velocity	$v_w = \omega/k_w = 2\pi\omega\lambda_w$
Wave number	$K_w = 2\pi/\lambda_w = \omega/v_w$
Group speed	$v_g = d\omega/dk_w$

STRING VIBRATION

A vibration in a string is a wave. Resonance causes a vibrating string to produce a sound with constant frequency, i.e. constant pitch. If the length or tension of the string is correctly adjusted, the sound produced is a musical tone. Vibrating strings are the basis of string instruments such as guitars, cellos, and pianos.

Wave

The velocity of propagation of a wave in a string (v) is proportional to the square root of the force of tension of the string (T) and inversely proportional to the square root of the linear density (ρ) of the string:

$$v = \sqrt{\frac{T}{\rho}}.$$

This relationship was discovered by Vincenzo Galilei in the late 1500s.

Derivation

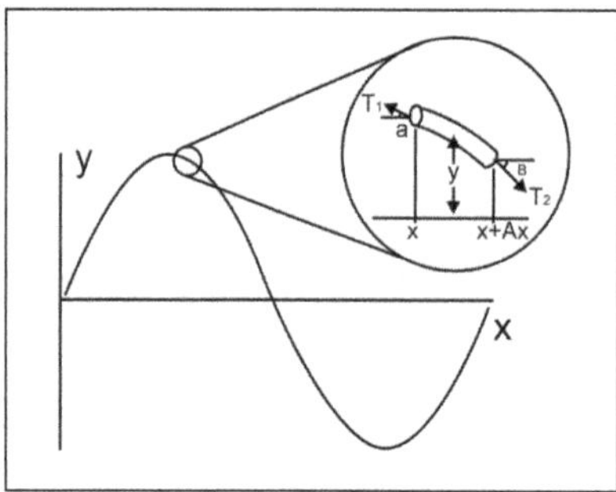

Let Δx be the length of a piece of string, m its mass, and ρ its linear density. If the horizontal component of tension in the string is a constant, T , then the tension acting on each side of the string segment is given by,

$$T_{1x} = T_1 \cos(\alpha) \approx T.$$

$$T_{2x} = T_2 \cos(\beta) \approx T.$$

If both angles are small, then the tensions on either side are equal and the net horizontal force is zero. From Newton's second law for the vertical component, the mass of this piece times its acceleration, a , will be equal to the net force on the piece:

$$\Sigma F_y = T_{1y} - T_{2y} = -T_2 \sin(\beta) + T_1 \sin(\alpha) = \Delta ma \approx \rho \Delta x \frac{\partial^2 y}{\partial t^2}.$$

Dividing this expression by T and substituting the first and second equations obtains,

$$\frac{\rho \Delta x}{T} \frac{\partial^2 y}{\partial t^2} = -\frac{T_2 \sin(\beta)}{T_2 \cos(\beta)} + \frac{T_1 \sin(\alpha)}{T_1 \cos(\alpha)} = -\tan(\beta) + \tan(\alpha)$$

The tangents of the angles at the ends of the string piece are equal to the slopes at the ends, with an additional minus sign due to the definition of alpha and beta. Using this fact and rearranging provides:

$$\frac{1}{\Delta x}\left(\left.\frac{\partial y}{\partial x}\right|^{x+\Delta x} - \left.\frac{\partial y}{\partial x}\right|^{x} \right) = \frac{\rho}{T} \frac{\partial^2 y}{\partial t^2}$$

In the limit that Δx approaches zero, the left hand side is the definition of the second derivative of y:

$$\frac{\partial^2 y}{\partial x^2} = \frac{\rho}{T}\frac{\partial^2 y}{\partial t^2}.$$

This is the wave equation for $y(x,t)$, and the coefficient of the second time derivative term is equal to v^{-2}; thus

$$v = \sqrt{\frac{T}{\rho}},$$

where v is the speed of propagation of the wave in the string. However, this derivation is only valid for vibrations of small amplitude; for those of large amplitude, Δx is not a good approximation for the length of the string piece, the horizontal component of tension is not necessarily constant, and the horizontal tensions are not well approximated by T.

Frequency of the Wave

Once the speed of propagation is known, the frequency of the sound produced by the string can be calculated. The speed of propagation of a wave is equal to the wavelength divided by the period λ, or multiplied by the frequency f:

$$v = \frac{\lambda}{\tau} = \lambda f.$$

If the length of the string is L, the fundamental harmonic is the one produced by the vibration whose nodes are the two ends of the string, so L is half of the wavelength of the fundamental harmonic. Hence one obtains Mersenne's laws:

$$f = \frac{v}{2L} = \frac{1}{2L}\sqrt{\frac{T}{\mu}}$$

where T is the tension (in Newtons), μ is the linear density (that is, the mass per unit length), and L is the length of the vibrating part of the string. Therefore:

- The shorter the string, the higher the frequency of the fundamental.
- The higher the tension, the higher the frequency of the fundamental.
- The lighter the string, the higher the frequency of the fundamental.

Moreover, if we take the nth harmonic as having a wavelength given by $\lambda_n = 2L/n$, then we easily get an expression for the frequency of the nth harmonic:

$$f_n = \frac{nv}{2L}$$

And for a string under a tension T with density μ, then,

$$f_n = \frac{n}{2L}\sqrt{\frac{T}{\mu}}$$

Observing String Vibrations

One can see the waveforms on a vibrating string if the frequency is low enough and the vibrating string is held in front of a CRT screen such as one of a television or a computer (*not* of an analog oscilloscope). This effect is called the stroboscopic effect, and the rate at which the string seems to vibrate is the difference between the frequency of the string and the refresh rate of the screen. The same can happen with a fluorescent lamp, at a rate that is the difference between the frequency of the string and the frequency of the alternating current. (If the refresh rate of the screen equals the frequency of the string or an integer multiple thereof, the string will appear still but deformed.) In daylight and other non-oscillating light sources, this effect does not occur and the string appears still but thicker, and lighter or blurred, due to persistence of vision.

A similar but more controllable effect can be obtained using a stroboscope. This device allows matching the frequency of the xenon flash lamp to the frequency of vibration of the string. In a dark room, this clearly shows the waveform. Otherwise, one can use bending or, perhaps more easily, by adjusting the machine heads, to obtain the same, or a multiple, of the AC frequency to achieve the same effect. For example, in the case of a guitar, the 6th (lowest pitched) string pressed to the third fret gives a G at 97.999 Hz. A slight adjustment can alter it to 100 Hz, exactly one octave above the alternating current frequency in Europe and most countries in Africa and Asia, 50 Hz. In most countries of the Americas—where the AC frequency is 60 Hz—altering A# on the fifth string, first fret from 116.54 Hz to 120 Hz produces a similar effect.

FREE VIBRATIONS IN SINGLE DEGREE-OF-FREEDOM SYSTEM

Free vibration (no external force) of a single degree-of-freedom system with viscous damping can be illustrated as,

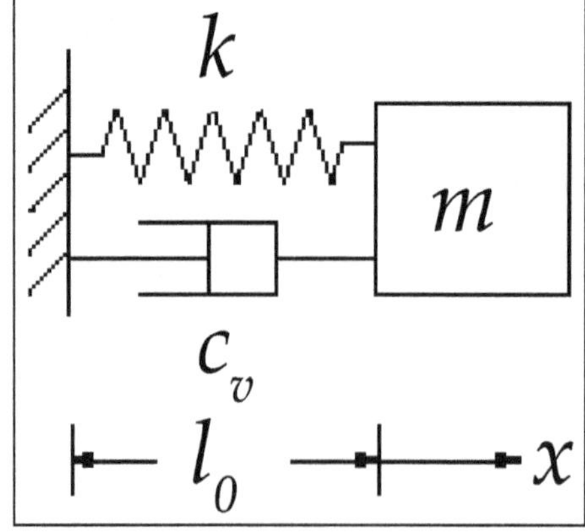

Damping that produces a damping force proportional to the mass's velocity is commonly referred to as "viscous damping", and is denoted graphically by a dashpot.

Time Solution for Damped SDOF Systems

For an unforced damped SDOF system, the general equation of motion becomes,

$$m\ddot{x} + c_v\dot{x} + kx = 0$$

with the initial conditions,

This equation of motion is a second order, homogeneous, ordinary differential equation (ODE). If all parameters (mass, spring stiffness, and viscous damping) are constants, the ODE becomes a linear ODE with constant coefficients and can be solved by the characteristic equation method. The characteristic equation for this problem is, $ms^2 + c_v s + k = 0$ which determines the two independent roots for the damped vibration problem. The roots to the characteristic equation fall into one of the following three cases:

1.	If $c_v^2 - 4mk$ < 0, the system is termed underdamped. The roots of the characteristic equation are complex conjugates, corresponding to oscillatory motion with an exponential decay in amplitude.
2.	If $c_v^2 - 4mk$ = 0, the system is termed critically-damped. The roots of the characteristic equation are repeated, corresponding to simple decaying motion with at most one overshoot of the system's resting position.
3.	If $c_v^2 - 4mk$ > 0, the system is termed overdamped. The roots of the characteristic equation are purely real and distinct, corresponding to simple exponentially decaying motion.

To simplify the solutions coming up, we define the critical damping c_c, the damping ratio ζ, and the damped vibration frequency ω_d as,

$$c_c = 2m\sqrt{\frac{k}{m}} = 2m\omega_n$$

$$\varsigma = \frac{c_v}{c_c}$$

$$\omega_d = \sqrt{1-\varsigma^2}\ \omega_n$$

where the natural frequency of the system ω_n is given by,

$$\omega_n = \sqrt{\frac{k}{m}}$$

Note that ω_d will equal ω_n when the damping of the system is zero (i.e. undamped). The time solutions for the free SDOF system is presented below for each of the three case scenarios.

Underdamped Systems

When $c_v^{\ 2} - 4mk$ < 0 (equivalent to ς < 1 or c_v < c_c), the characteristic equation has a pair of complex conjugate roots. The displacement solution for this kind of system is,

$$x(t)=c_1 e^{\left(-\varsigma+i\sqrt{1-\varsigma^2}\right)\omega_n t}+c_2 e^{\left(-\varsigma-i\sqrt{1-\varsigma^2}\right)\omega_n t}$$

$$=e^{-\varsigma\omega_n t}\left[d_1\cos(\omega_d t)+d_2\sin(\omega_d t)\right]$$

$$\Rightarrow x(t)=\underbrace{e^{-\varsigma\omega_n t}}_{\text{Exponentially decay}}\underbrace{\left[x_0\cos(\omega_d t)+\frac{v_0+\varsigma\omega_n x_0}{\omega_d}\sin(\omega_d t)\right]}_{\text{Periodic motion}}$$

An alternate but equivalent solution is given by,

$$x(t)=\mathrm{A}_0\underbrace{e^{-\varsigma\omega_n t}}_{\text{Exponentially decay}}\underbrace{\cos\left(\omega_d t-\varphi_0\right)}_{\text{Periodic}}$$

The displacement plot of an underdamped system would appear as,

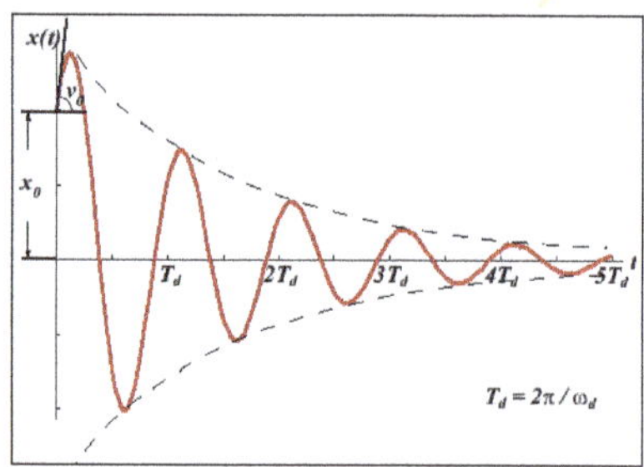

Note that the displacement amplitude decays exponentially (i.e. the natural logarithm of the amplitude ratio for any two displacements separated in time by a constant ratio is a constant; long-winded),

$$\frac{A_k}{A_{k+1}}=\frac{A_0 e^{-\varsigma\omega_n(kT_d)}\cos(\varphi_0)}{A_0 e^{-\varsigma\omega_n[(k+1)T_d]}\cos(\varphi_0)}=\frac{e^{-\varsigma\omega_n(kT_d)}}{e^{-\varsigma\omega_n[(k+1)T_d]}}=e^{\varsigma\omega_n T_d}$$

$$\Rightarrow \mathrm{In}\left(\frac{A_k}{A_{k+1}}\right)=\varsigma\omega_n T_d=\varsigma\omega_n\frac{2\pi}{\omega_d}=\frac{2\pi\varsigma}{\sqrt{1-\varsigma^2}}$$

where $T_d=\dfrac{1}{f_d}=\dfrac{2\pi}{\omega_d}$ is the period of the damped vibration.

Critically-Damped Systems

When $c_v^{\ 2} - 4mk$ = 0 (equivalent to ς = 1 or c_v = c_c), the characteristic equation has repeated real roots. The displacement solution for this kind of system is,

$$x(t)=(c_1+c_2 t)e^{-\omega_n t}$$

$$\Rightarrow x(t)=e^{-\omega_n t}\left[x_0+(v_{0+}\omega_n x_0)t\right]$$

The critical damping factor c_c can be interpreted as the minimum damping that result in non-periodic motion (i.e. simple decay).

The displacement plot of a critically-damped system with positive initial displacement and velocity would appear as,

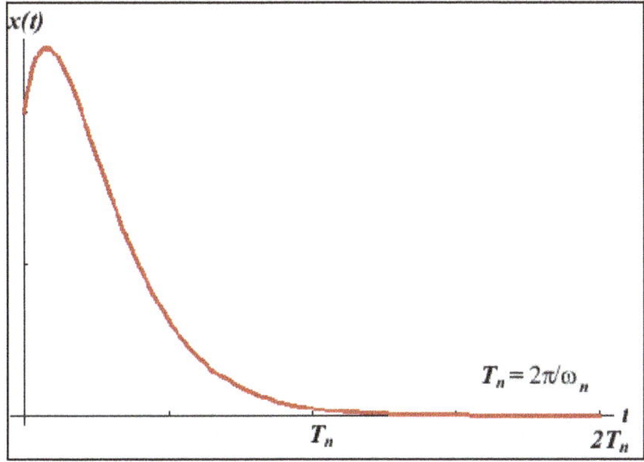

The displacement decays to a negligible level after one natural period, T_n. Note that if the initial velocity v_o is negative while the initial displacement x_o is positive, there will exist one overshoot of the resting position in the displacement plot.

Overdamped Systems

When $cv^2 - 4mk > 0$ (equivalent to $\varsigma > 1$ or $c_v > c_c$), the characteristic equation has two distinct real roots. The displacement solution for this kind of system is,

$$x(t) = c_1 e^{\left(-\varsigma+\sqrt{\varsigma^2-1}\right)\omega_n t} + c_2 e^{\left(-\varsigma-\sqrt{\varsigma^2-1}\right)\omega_n t}$$

$$\Rightarrow x(t) = \frac{x_0\omega_n\left(\varsigma+\sqrt{\varsigma^2-1}\right)+v_0}{2\omega_n\sqrt{\varsigma^2-1}} e^{\left(-\varsigma+\sqrt{\varsigma^2-1}\right)\omega_n t} + \frac{-x_0\omega_n\left(\varsigma-\sqrt{\varsigma^2-1}\right)-v_0}{2\omega_n\sqrt{\varsigma^2-1}} e^{\left(-\varsigma-\sqrt{\varsigma^2-1}\right)\omega_n t}$$

The displacement plot of an overdamped system would appear as,

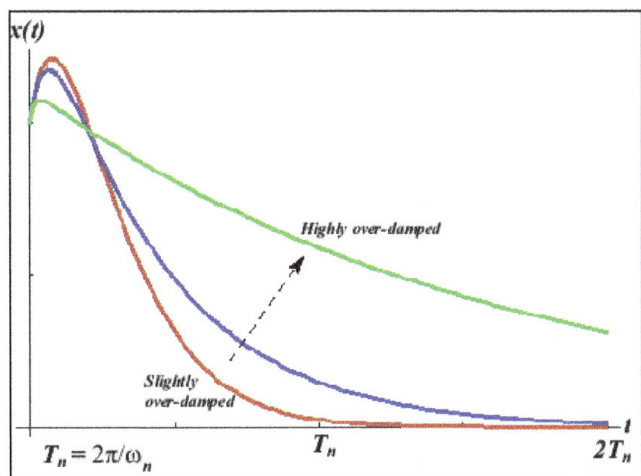

The motion of an overdamped system is non-periodic, regardless of the initial conditions. The larger the damping, the longer the time to decay from an initial disturbance.

If the system is heavily damped, $\zeta \gg 1$, the displacement solution takes the approximate form,

$$x(t) \approx x_0 + \frac{v_0}{2\zeta\omega_n}\left(1 - e^{-2\zeta\omega_n t}\right)$$

FORCED VIBRATIONS

Undamped and Forced Vibrations

The differential equation in this case is:

$$mu'' + ku = F(t)$$

This is just a nonhomogeneous differential equation and we know how to solve these. The general solution will be,

$$u(t) = u_c(t) + U_P(t)$$

Where, the complementary solution is the solution to the free, undamped vibration case. To get the particular solution we can use either undetermined coefficients or variation of parameters depending on which we find easier for a given forcing function.

There is a particular type of forcing function that we should take a look at since it leads to some interesting results. Let's suppose that the forcing function is a simple periodic function of the form:

$$F(t) = F_0 \cos(\omega t) \quad \text{OR} \quad F(t) = F_0 \sin(\omega t)$$

For the purposes of this discussion we'll use the first one. Using this, the IVP becomes,

$$mu'' + ku = F_0 \cos(\omega t)$$

The complementary solution, as pointed out above, is just,

$$u_c(t) = c_1 \cos(\omega_0 t) + c_2 \sin(\omega_0 t)$$

where, ω_0 is the natural frequency.

We will need to be careful in finding a particular solution. The reason for this will be clear if we use undetermined coefficients. With undetermined coefficients our guess for the form of the particular solution would be,

$$U_P(t) = A\cos(\omega t) B \sin(\omega t)$$

Now, this guess will be problems if $\omega_0=\omega$. If this were to happen the guess for the particular solution is exactly the complementary solution and so we'd need to add in a t. Of course, if we don't have ωo=ω then there will be nothing wrong with the guess.

So, we will need to look at this in two cases.

$$1.\omega_0 \neq \omega$$

In this case our initial guess is okay since it won't be the complementary solution. Upon differentiating the guess and plugging it into the differential equation and simplifying we get,

$$(-m\omega^2 A + kA)\cos(\omega t) + (-m\omega^2 B + kB)\sin(\omega t) = F_0\cos(\omega t)$$

Setting coefficients equal gives us,

$$\begin{aligned} \cos(\omega t): &\quad (-m\omega^2 + k)A = F_0 \;\Rightarrow\; A = \frac{F_0}{k - m\omega^2} \\ \sin(\omega t): &\quad (-m\omega^2 + k)B = 0 \;\Rightarrow\; B = 0 \end{aligned}$$

The particular solution is then,

$$\begin{aligned} U_P(t) &= \frac{F_0}{k - m\omega^2}\cos(\omega t) \\ &= \frac{F_0}{m\left(\frac{k}{m} - \omega^2\right)}\cos(\omega t) \\ &= \frac{F_0}{m\left(\omega_0^2 - \omega^2\right)}\cos(\omega t) \end{aligned}$$

Note that we rearranged things a little. Depending on the form that you'd like the displacement to be in we can have either of the following:

$$u(t) = c_1\cos(\omega_0 t) + c_2\sin(\omega_0 t) + \frac{F_0}{m(\omega_0^2 - \omega^2)}\cos(\omega t)$$

$$u(t) = R\cos(\omega_0 t - \delta) + \frac{F_0}{m(\omega_0^2 - \omega^2)}\cos(\omega t)$$

If we used the sine form of the forcing function we could get a similar formula.

$$2.\omega_0 = \omega$$

In this case we will need to add in a tt to the guess for the particular solution.

$$U_P(t) = At\cos(\omega_0 t) + Bt\sin(\omega_0 t)$$

Note that we went ahead and acknowledge that $\omega o = \omega_0 = \omega$ in our guess. Differentiating our guess, plugging it into the differential equation and simplifying gives us the following:

$$\begin{aligned} &(-m\omega_0^2 + k)At\cos(\omega t) + (-m\omega_0^2 + k)Bt\sin(\omega t) + \\ &\qquad 2m\omega_0 B\cos(\omega t) - 2m\omega_0 A\sin(\omega t) = F_0\cos(\omega t) \end{aligned}$$

Before setting coefficients equal, let's remember the definition of the natural frequency and note that,

$$-m\omega_0^2 + k = -m\left(\sqrt{\frac{k}{m}}\right)^2 + k = -m\left(\frac{k}{m}\right) + k = 0$$

So, the first two terms actually drop out (which is a very good thing) and this gives us,

$$2m\omega_0 B\cos(\omega t) - 2m\omega_0 A\sin(\omega t) = F_0\cos(\omega t)$$

Now let's set coefficient equal:

$$\cos(\omega t): \quad 2m\omega_0 B = F_0 \quad \Rightarrow B = \frac{F_0}{2m\omega_0}$$

$$\sin(\omega t): \quad 2m\omega_0 A = 0 \quad \Rightarrow A = 0$$

In this case the particular will be,

$$U_P(t) = \frac{F_0}{2m\omega_0} t\sin(\omega_0 t)$$

The displacement for this case is then,

$$u(t) = c_1\cos(\omega_0 t) + c_2\sin(\omega_0 t) + \frac{F_0}{2m\omega_0} t\sin(\omega_0 t)$$

$$u(t) = R\cos(\omega_0 t - \delta) + \frac{F_0}{2m\omega_0} t\sin(\omega_0 t)$$

depending on the form that you prefer for the displacement.

So, what was the point of the two cases here? Well in the first case, $\omega_0 \neq \omega$ our displacement function consists of two cosines and is nice and well behaved for all time.

In contrast, the second case, $\omega_0 = \omega$ will have some serious issues at t increases. The addition of the t in the particular solution will mean that we are going to see an oscillation that grows in amplitude as t increases. This case is called resonance and we would generally like to avoid this at all costs.

In this case resonance arose by assuming that the forcing function was,

$$F(t) = F_0\cos(\omega_0 t)$$

We would also have the possibility of resonance if we assumed a forcing function of the form:

$$F(t) = F_0\sin(\omega_0 t)$$

We should also take care to not assume that a forcing function will be in one of these two forms. Forcing functions can come in a wide variety of forms. If we do run into a forcing function different from the one that used here you will have to go through undetermined coefficients or variation of parameters to determine the particular solution.

Forced and Damped Vibrations

This is the full blown case where we consider every last possible force that can act upon the system. The differential equation for this case is,

$$mu'' + \gamma u' + ku = F(t)$$

The displacement function this time will be,

$$u(t) = u_c(t) + U_P(t)$$

Where, the complementary solution will be the solution to the free, damped case and the particular solution will be found using undetermined coefficients or variation of parameter, whichever is most convenient to use.

There are a couple of things to note here about this case. First, from our work back in the free, damped case we know that the complementary solution will approach zero as t increases. Because of this the complementary solution is often called the transient solution in this case.

Also, because of this behavior the displacement will start to look more and more like the particular solution as t increases and so the particular solution is often called the steady state solution or forced response.

TORSIONAL VIBRATION

Torsional vibration is angular vibration of an object—commonly a shaft along its axis of rotation. Torsional vibration is often a concern in power transmission systems using rotating shafts or couplings where it can cause failures if not controlled. A second effect of torsional vibrations applies to passenger cars. Torsional vibrations can lead to seat vibrations or noise at certain speeds. Both reduce the comfort.

In ideal power generation, or transmission, systems using rotating parts, not only the torques applied or reacted are "smooth" leading to constant speeds, but also the rotating plane where the power is generated (or input) and the plane it is taken out (output) are the same. In reality this is not the case. The torques generated may not be smooth (e.g., internal combustion engines) or the component being driven may not react to the torque smoothly (e.g., reciprocating compressors), and the power generating plane is normally at some distance to the power takeoff plane. Also, the components transmitting the torque can generate non-smooth or alternating torques (e.g., elastic drive belts, worn gears, misaligned shafts). Because no material can be infinitely stiff, these alternating torques applied at some distance on a shaft cause twisting vibration about the axis of rotation.

Sources of Torsional Vibration

Torsional vibration can be introduced into a drive train by the power source. But even a drive train with a very smooth rotational input can develop torsional vibrations through internal components. Common sources are:

- Internal combustion engine: The torsional vibrations of the not continuous combustion and the crank shaft geometry itself cause torsional vibrations.
- Reciprocating compressor: The pistons experience discontinuous forces from the compression.
- Universal joint: The geometry of this joint causes torsional vibrations if the shafts are not parallel.
- Stick slip: During the engagement of a friction element, stick slip situations create torsional vibrations.
- Lash: Drive train lash can cause torsional vibrations if the direction of rotation is changed or if the flow of power, i.e. driver vs. driven, is reversed.

Crankshaft Torsional Vibration

Torsional vibration is a concern in the crankshafts of internal combustion engines because it could break the crankshaft itself; shear-off the flywheel; or cause driven belts, gears and attached components to fail, especially when the frequency of the vibration matches the torsional resonant frequency of the crankshaft. Causes of the torsional vibration are attributed to several factors.

- Alternating torques are generated by the slider-crank mechanism of the crankshaft, connecting rod, and piston.
    - The cylinder pressure due to combustion is not constant through the combustion cycle.
    - The slider-crank mechanism does not output a smooth torque even if the pressure is constant (e.g., at top dead centre there is no torque generated).
    - The motion of the piston mass and connecting rod mass generate alternating torques often referred to as "inertia" torques.
- Engines with six or more cylinders in a straight line configuration can have very flexible crankshafts due to their long length.
- 2 Stroke Engines generally have smaller bearing overlap between the main and the pin bearings due to the larger stroke length, hence increasing the flexibility of the Crankshaft due to decreased stiffness.
- There is inherently little damping in a crankshaft to reduce the vibration except for the shearing resistance of oil film in the main and conrod bearings.

If torsional vibration is not controlled in a crankshaft it can cause failure of the crankshaft or any accessories that are being driven by the crankshaft (typically at the front of the engine; the inertia of the flywheel normally reduces the motion at the rear of the engine).

This potentially damaging vibration is often controlled by a torsional damper that is located at the front nose of the crankshaft (in automobiles it is often integrated into the front pulley). There are two main types of torsional dampers.

- Viscous dampers consist of an inertia ring in a viscous fluid. The torsional vibration of the crankshaft forces the fluid through narrow passages that dissipates the vibration as heat. The viscous torsional damper is analogous to the hydraulic shock absorber in a car's suspension.
- Tuned absorber type of "dampers" often referred to as a harmonic dampers or harmonic balancers (even though it technically does not damp or balance the crankshaft). This damper uses a spring element (often rubber in automobile engines) and an inertia ring that is typically tuned to the first torsional natural frequency of the crankshaft. This type of damper reduces the vibration at specific engine speeds when an excitation torque excites the first natural frequency of the crankshaft, but not at other speeds. This type of damper is analogous to the tuned mass dampers used in skyscrapers to reduce the building motion during an earthquake.

Torsional Vibrations in Electromechanical Drive Systems

Torsional vibrations of drive systems usually result in a significant fluctuation of the rotational speed of the rotor of the driving electric motor. Such oscillations of the angular speed superimposed on the average rotor rotational speed cause more or less severe perturbation of the electromagnetic flux and thus additional oscillations of the electric currents in the motor windings. Then, the generated electromagnetic torque is also characterized by additional variable in time components which induce torsional vibrations of the drive system. According to the above, mechanical vibrations of the drive system become coupled with the electrical vibrations of currents in the motor windings. Such a coupling is often complicated in character and thus computationally troublesome. Because of this reason, till present majority of authors used to simplify the matter regarding mechanical vibrations of drive systems and electric current vibrations in the motor windings as mutually uncoupled. Then, the mechanical engineers applied the electromagnetic torques generated by the electric motors as 'a priori' assumed excitation functions of time or of the rotor-to-stator slip, e.g. in paper usually basing on numerous experimental measurements carried out for the given electric motor dynamic behaviours. For this purpose, by means of measurement results, proper approximate formulas have been developed, which describe respective electromagnetic external excitations produced by the electric motor. However, the electricians thoroughly modelled electric current flows in the electric motor windings, but they usually reduced the mechanical drive system to one or seldom to at most a few rotating rigid bodies, as e.g. in In many cases, such simplifications yield sufficiently useful results for engineering applications, but very often they can lead to remarkable inaccuracies, since many qualitative dynamic properties of the mechanical systems, e.g. their mass distribution, torsional flexibility and damping effects, are being neglected. Thus, an influence of drive system vibratory behaviour on the electric machine rotor angular speed fluctuation, and in this way on the electric current oscillations in the rotor and stator windings, can not be investigated with a satisfactory precision.

Mechanical vibrations and deformations are phenomena associated with an operation of majority of railway vehicle drivetrain structures. The knowledge about torsional vibrations in transmission systems of railway vehicles is of a great importance in the fields dynamics of mechanical systems. Torsional vibrations in the railway vehicle drive train are generated by several phenomena. Generally, these phenomena are very complex and they can be divided into two main parts.

- To the first one belongs the electromechanical interaction between of the railway drive system including the: electric motor, gears, the driven part of disc clutch and driving parts of the gear clutch.
- To the second one belong torsional vibrations of the flexible wheels, and wheelsets caused by variation of adhesion forces in the wheel-rail contact zone.

An interaction of the adhesion forces has nonlinear features which are related to the creep value and strongly depends on the wheel-rail zone condition and track geometry (when driving on a curve section of the track). In many modern mechanical systems torsional structural deformability plays an important role. Often the study of railway vehicle dynamics using the rigid multibody methods without torsionally deformable elements are used. This approach does not allow to analyse self-excited vibrations which have an important influence on the wheel-rail longitudinal interaction. A dynamic modelling of the electrical drive systems coupled with elements of a driven machine or vehicle is particularly important when the purpose of such modelling is to obtain an information about the transient phenomena of system operation, like a run-up, run-down and loss of adhesion in the wheel-rail zone. The modelling of an electromechanical interaction between the electric driving motor and the machine as well as to an influence of the self-excited torsional vibrations in the drive system.

Measuring Torsional Vibration on Physical Systems

The most common way to measure torsional vibration is the approach of using equidistant pulses over one shaft revolution. Dedicated shaft encoders as well as gear tooth pickup transducers (induction, hall-effect, variable reluctance,etc.) can generate these pulses. The resulting encoder pulse train is converted into either a digital rpm reading or a voltage proportional to the rpm.

The use of a dual-beam laser is another technique that is used to measure torsional vibrations. The operation of the dual-beam laser is based on the difference in reflection frequency of two perfectly aligned beams pointing at different points on a shaft. Despite its specific advantages, this method yields a limited frequency range, requires line-of-sight from the part to the laser, and represents multiple lasers in case several measurement points need to be measured in parallel.

Torsional Vibration Software

There are many software packages that are capable of solving the torsional vibration system of equations. Torsional vibration specific codes are more versatile for design and system validation purposes and can produce simulation data that can readily compared to published industry standards. These codes make it easy to add system branches, mass-elastic data, steady-state loads, transient disturbances and many other items only a rotordynamicist would need. Torsional vibration specific codes:

- ARMD TORSION (Rotor Bearing Technology & Software, Inc.) - Commercial FEA-based software for performing damped and undamped torsional natural frequencies, mode shapes, steady-state and time-transient response of mechanical drive trains with inputs of various types of external excitation, synchronous motor start-up torque, compressor torques, and electrical system disturbances. Used worldwide by researchers, OEMs and end-users across all industries.

MECHANICAL RESONANCE

Mechanical resonance is the tendency of a mechanical system to respond at greater amplitude when the frequency of its oscillations matches the system's natural frequency of vibration (its *resonance frequency* or *resonant frequency*) than it does at other frequencies. It may cause violent swaying motions and even catastrophic failure in improperly constructed structures including bridges, buildings and airplanes. This is a phenomenon known as resonance disaster.

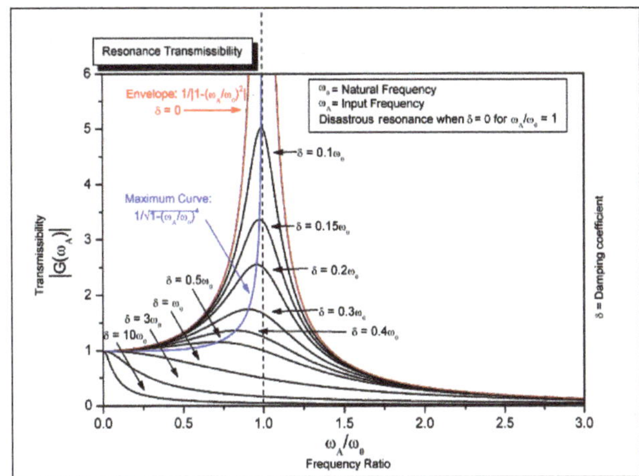

Graph showing mechanical resonance in a mechanical oscillatory system.

Avoiding resonance disasters is a major concern in every building, tower and bridge construction project. The Taipei 101 building relies on a 660-ton pendulum—a tuned mass damper—to modify the response at resonance. Furthermore, the structure is designed to resonate at a frequency which does not typically occur. Buildings in seismic zones are often constructed to take into account the oscillating frequencies of expected ground motion. In addition, engineers designing objects having engines must ensure that the mechanical resonant frequencies of the component parts do not match driving vibrational frequencies of the motors or other strongly oscillating parts.

Many resonant objects have more than one resonance frequency. It will vibrate easily at those frequencies, and less so at other frequencies. Many clocks keep time by mechanical resonance in a balance wheel, pendulum, or quartz crystal.

The natural frequency of a simple mechanical system consisting of a weight suspended by a spring is:

$$f = \frac{1}{2\pi}\sqrt{\frac{k}{m}}$$

where m is the mass and k is the spring constant.

A swing set is a simple example of a resonant system with which most people have practical experience. It is a form of pendulum. If the system is excited (pushed) with a period between pushes equal to the inverse of the pendulum's natural frequency, the swing will swing higher and higher, but if excited at a different frequency, it will be difficult to move. The resonance frequency of a pendulum, the only frequency at which it will vibrate, is given approximately, for small displacements, by the equation:

$$f = \frac{1}{2\pi}\sqrt{\frac{g}{L}}$$

where g is the acceleration due to gravity (about 9.8 m/s^2 near the surface of Earth), and L is the length from the pivot point to the center of mass.(An elliptic integral yields a description for any displacement). Note that, in this approximation, the frequency does not depend on mass.

Mechanical resonators work by transferring energy repeatedly from kinetic to potential form and back again. In the pendulum, for example, all the energy is stored as gravitational energy (a form of potential energy) when the bob is instantaneously motionless at the top of its swing. This energy is proportional to both the mass of the bob and its height above the lowest point. As the bob descends and picks up speed, its potential energy is gradually converted to kinetic energy (energy of movement), which is proportional to the bob's mass and to the square of its speed. When the bob is at the bottom of its travel, it has maximum kinetic energy and minimum potential energy. The same process then happens in reverse as the bob climbs towards the top of its swing.

Some resonant objects have more than one resonance frequency, particularly at harmonics (multiples) of the strongest resonance. It will vibrate easily at those frequencies, and less so at other frequencies. It will "pick out" its resonance frequency from a complex excitation, such as an impulse or a wideband noise excitation. In effect, it is filtering out all frequencies other than its resonance. In the example above, the swing cannot easily be excited by harmonic frequencies, but can be excited by subharmonics.

Examples:

Resonance Rings exhibit at California Science Center.

Various examples of mechanical resonance include:

- Musical instruments (acoustic resonance).

- Most clocks keep time by mechanical resonance in a balance wheel, pendulum, or quartz crystal.
- Tidal resonance of the Bay of Fundy.
- Orbital resonance as in some moons of the solar system's gas giants.
- The resonance of the basilar membrane in the ear.
- A wineglass breaking when someone sings a loud note at exactly the right pitch.

Resonance may cause violent swaying motions in improperly constructed structures, such as bridges and buildings. The London Millennium Footbridge (nicknamed the *Wobbly Bridge*) exhibited this problem. A faulty bridge can even be destroyed by its resonance. Mechanical systems store potential energy in different forms. For example, a spring/mass system stores energy as tension in the spring, which is ultimately stored as the energy of bonds between atoms.

Resonance Disaster

In mechanics and construction a resonance disaster describes the destruction of a building or a technical mechanism by induced vibrations at a system's resonance frequency, which causes it to oscillate. Periodic excitation optimally transfers to the system the energy of the vibration and stores it there. Because of this repeated storage and additional energy input the system swings ever more strongly, until its load limit is exceeded.

Tacoma Narrows Bridge

The dramatic, rhythmic twisting that resulted in the 1940 collapse of "Galloping Gertie", the original Tacoma Narrows Bridge, is sometimes characterized in physics textbooks as a classic example of resonance. The catastrophic vibrations that destroyed the bridge were due to an oscillation caused by interactions between the bridge and the winds passing through its structure—a phenomenon known as aeroelastic flutter.

Other Examples:

- Collapse of Broughton Suspension Bridge (due to soldiers walking in step).
- Collapse of Angers Bridge.
- Collapse of Königs Wusterhausen Central Tower.
- Resonance of the Millennium Bridge.

Applications

Various method of inducing mechanical resonance in a medium exist. Mechanical waves can be generated in a medium by subjecting an electromechanical element to an alternating electric field having a frequency which induces mechanical resonance and is below any electrical resonance frequency. Such devices can apply mechanical energy from an external source to an element to mechanically stress the element or apply mechanical energy produced by the element to an external load.

The United States Patent Office classifies devices that tests mechanical resonance under subclass 579, resonance, frequency, or amplitude study, of Class 73, Measuring and testing. This subclass is itself indented under subclass 570, Vibration. Such devices test an article or mechanism by subjecting it to a vibratory force for determining qualities, characteristics, or conditions thereof, or sensing, studying or making analysis of the vibrations otherwise generated in or existing in the article or mechanism. Devices include right methods to cause vibrations at a natural mechanical resonance and measure the frequency and/or amplitude the resonance made. Various devices study the amplitude response over a frequency range is made. This includes nodal points, wave lengths, and standing wave characteristics measured under predetermined vibration conditions.

HARMONIC OSCILLATOR

The harmonic oscillator is a model which has several important applications in both classical and quantum mechanics. It serves as a prototype in the mathematical treatment of such diverse phenomena as elasticity, acoustics, AC circuits, molecular and crystal vibrations, electromagnetic fields and optical properties of matter.

Classical Oscillator

A simple realization of the harmonic oscillator in classical mechanics is a particle which is acted upon by a restoring force proportional to its displacement from its equilibrium position. Considering motion in one dimension, this means:

$$F = -kx$$

Such a force might originate from a spring which obeys Hooke's law, According to Hooke's law, which applies to real springs for sufficiently small displacements, the restoring force is proportional to the displacement—either stretching or compression—from the equilibrium position.

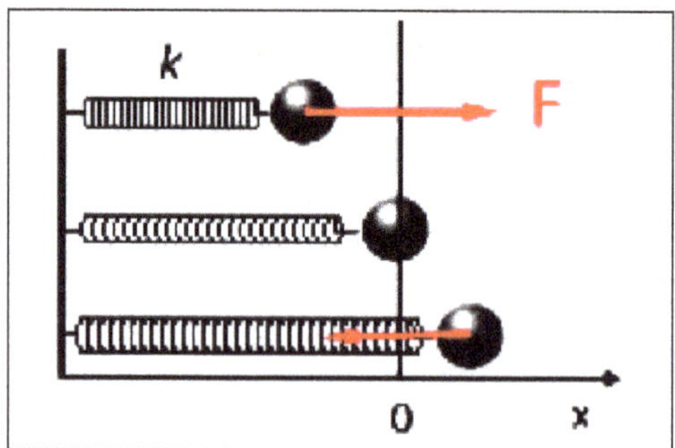

Spring obeying Hooke's law.

The *force constant* k is a measure of the stiffness of the spring. The variable x is chosen equal to zero at the equilibrium position, positive for stretching, negative for compression. The negative sign in Equation $F = -kx$ reflects the fact that F is a *restoring* force, always in the opposite sense to the displacement x.

Applying Newton's second law to the force from Equation $F = -kx$, we find x.

$$F = m\frac{d^2x}{dx^2} = -kx$$

where m is the mass of the body attached to the spring, which is itself assumed massless. This leads to a differential equation of familiar form, although with different variables:

$$\ddot{x}(t) + \omega^2 x(t) = 0$$

With $\omega^2 \equiv \frac{k}{m}$

The dot notation (introduced by Newton himself) is used in place of primes when the independent variable is time. The general solution to Equation $\ddot{x}(t) + \omega^2 x(t) = 0$ is,

$$x(t) = A\sin\omega t + B\cos\omega t$$

which represents periodic motion with a sinusoidal time dependence. This is known as simple harmonic motion and the corresponding system is known as a harmonic oscillator. The oscillation occurs with a constant angular frequency,

$$\omega = \sqrt{\frac{k}{m}} \text{ radians per second}$$

This is called the natural frequency of the oscillator. The corresponding circular (or angular) frequency in Hertz (cycles per second) is,

$$\nu = \frac{\omega}{2\pi} = \frac{1}{2}\pi\sqrt{\frac{k}{m}} Hz$$

The general relation between force and potential energy in a conservative system in one dimension is,

$$F = \frac{-dV}{dx}$$

Thus the potential energy of a harmonic oscillator is given by,

$$V(x) = \frac{1}{2}kx^2$$

which has the shape of a parabola, as drawn in figure. A simple computation shows that the oscillator moves between positive and negative turning points $\pm x_{max}$ where the total energy EE equals the potential energy $\frac{1}{2}kx^2{}_{max}$ while the kinetic energy is momentarily zero. In contrast, when the oscillator moves past $x = 0$, the kinetic energy reaches its maximum value while the potential energy equals zero.

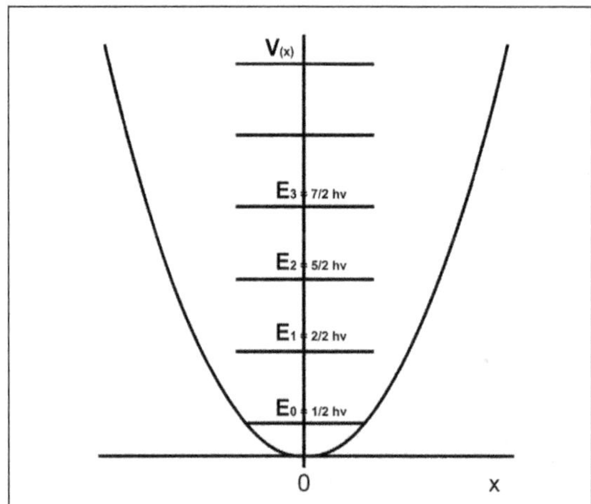

Potential energy function and first few energy levels for harmonic oscillator.

Harmonic Oscillator in Quantum Mechanics

Given the potential energy in Equation $V(x)=\frac{1}{2}kx^2$, we can write down the Schrödinger equation for the one-dimensional harmonic oscillator:

$$-\frac{\hbar^2}{2m}\psi''(x)+\frac{1}{2}kx^2\psi(x)=E\psi(x)$$

For the first time we encounter a differential equation with *non-constant* coefficients, which is a much greater challenge to solve. We can combine the constants in equation $-\frac{\hbar^2}{2m}\psi''(x)+\frac{1}{2}kx^2\psi(x)=E\psi(x)$ to two parameters,

$$\alpha^2=\frac{mk}{\hbar^2}$$

and

$$\lambda=\frac{2mE}{\hbar^2\alpha}$$

and redefine the independent variable as,

$$\xi=\alpha^{1/2}x$$

This reduces the Schrödinger equation to,

$$\psi''(\xi)+(\lambda-\xi^2)\psi(\xi)=0$$

The range of the variable x (also ξ) must be taken from $-\infty$ to $+\infty$, there being no finite cutoff as in the case of the particle in a box. A useful first step is to determine the asymptotic solution to

Equation $\xi = \alpha^{1/2}x$, that is, the form of ψ(ξ) as ξ→±∞. For sufficiently large values of |ξ|, ξ2≫λ and the differential equation is approximated by,

$$\psi''(\xi) - \xi^2\psi(\xi) \approx 0$$

This suggests the following manipulation:

$$\left(\frac{d^2}{d\xi^2} - \xi^2\right)\psi(\xi) \approx \left(\frac{d}{d\xi} - \xi\right)\left(\frac{d}{d\xi} + \xi\right)\psi(\xi) \approx 0$$

The first-order differential equation,

$$\psi'(\xi) + \xi\psi(\xi) = 0$$

can be solved exactly to give,

$$\psi(\xi) = const.e^{-\xi^2/2}$$

Remarkably, this turns out to be an exact solution of the Schrödinger equation $\psi''(\xi) + (\lambda - \xi^2)\psi(\xi) = 0$ with λ=1. Using Equation $\lambda = \dfrac{2mE}{\hbar^2\alpha}$, this corresponds to an energy,

$$E = \frac{\lambda\hbar^2\alpha}{2m} = \frac{1}{2}\hbar\sqrt{\frac{k}{m}} = \frac{1}{2}\hbar\omega$$

where ω is the natural frequency of the oscillator according to classical mechanics. The function in Equation $\psi(\xi) = const.e^{-\xi^2/2}$ has the form of a Gaussian, the bell-shaped curve so beloved in the social sciences. The function has no nodes, which leads us to conclude that this represents the ground state of the system.The ground state is usually designated with the quantum number n=0 (the particle in a box is a exception, with *n*=1 labeling the ground state). Reverting to the original variable *x*, we write,

$$\psi_0(x) = conste^{-\alpha x^2/2}$$

With,

$$\alpha = (mk / \hbar^2)^{1/2}$$

With help of the well-known definite integral,

$$\int_{-\infty}^{\infty} e^{-\alpha x^2}dx = \sqrt{\frac{\pi}{\alpha}}$$

we find the normalized eigenfunction,

$$\psi_0(x) = (\frac{\alpha}{\pi})^{1/4}e^{-\alpha x^2/2}$$

with the corresponding eigenvalue,

$$E_0 = \frac{1}{2}\hbar\omega$$

Drawing from our experience with the particle in a box, we might surmise that the first excited state of the harmonic oscillator would be a function similar to equation $\psi_0(x) = (\frac{\alpha}{\pi})^{1/4} e^{-\alpha x^2/2}$, but with a node at x=0, say,

$$\psi 1(x) = const x e^{-\alpha x^2/2}$$

This is orthogonal to $\psi_0(x)$ by symmetry and is indeed an eigenfunction with the eigenvalue,

$$E_1 = \frac{3}{2}\hbar\omega$$

Continuing the process, we try a function with two nodes,

$$\psi_2 = const(x^2 - a)e^{-\alpha x^2/2}$$

In Gaussian Integrals, we determine that with $a = \frac{1}{2}$ makes $\psi_2(x)$ orthogonal to $\psi_0(x)$ and $\psi_1(x)$. We verify that this is another eigenfunction, corresponding to,

$$E_2 = \frac{5}{2}\hbar\omega$$

The general result, which follows from a more advanced mathematical analysis, gives the following formula for the normalized eigenfunctions:

$$\psi_n(x) = (\frac{\sqrt{\alpha}}{2^n n! \sqrt{\pi}})^{1/2} H_n(\sqrt{\alpha}x)e^{-\alpha x^2/2}$$

where $H_n(\xi)$ represents the Hermite polynomial of degree n. The first few Hermite polynomials are,

$$H_0(\xi) = 1$$

$$H_1(\xi) = 2\xi$$

$$H_2(\xi) = 4\xi^2 - 2$$

$$H_3(\xi) = 8\xi^3 - 12\xi$$

The four lowest harmonic-oscillator eigenfunctions are plotted in figure. Note the topological resemblance to the corresponding particle-in-a-box eigenfunctions.

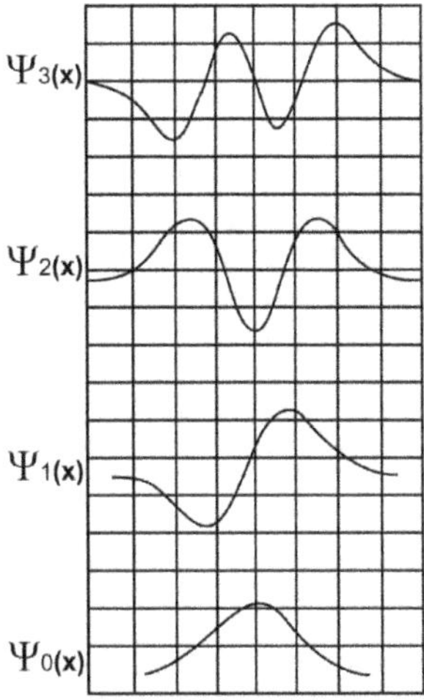

Harmonic oscillator eigenfunctions for n=0, 1, 2, 3.

The eigenvalues are given by the simple formula,

$$E_n = \left(n + \frac{1}{2}\right)\hbar\omega$$

The ground-state energy $E_0 = \frac{1}{2}\hbar\omega$ is greater than the classical value of zero, again a consequence of the uncertainty principle. This means that the oscillator is always oscillating.

It is remarkable that the difference between successive energy eigenvalues has a constant value,

$$\Delta E = E_{n+1} - E_n = \hbar\omega = h\nu$$

This is reminiscent of Planck's formula for the energy of a photon. It comes as no surprise then that the quantum theory of radiation has the structure of an assembly of oscillators, with each oscillator representing a mode of electromagnetic waves of a specified frequency.

References

- Molteno, T. C. A.; N. B. Tufillaro (September 2004). "An experimental investigation into the dynamics of a string". American Journal of Physics. 72 (9): 1157–1169. Bibcode:2004AmJPh..72.1157M. doi:10.1119/1.1764557
- Vibration-theory: petrowiki.org, Retrieved 30 April, 2019
- Feese, Hill. "Prevention of Torsional Vibration Problems in Reciprocating Machinery" (PDF). Engineering Dynamics Incorporated. Retrieved 17 October, 2013
- Vibrations, Classes, DE: tutorial.math.lamar.edu, Retrieved 15 June, 2019

5

Mechanics of Materials

Mechanics of materials is a field of study which deals with the behavior of solid objects as they are subjected to stresses and strains. Stress–strain curve, yield strength and compressive strength are some of the important concepts which are dealt with in this field. Mechanics of materials is a vast area that branches out into these significant sub-disciplines which have been thoroughly discussed in this chapter.

STRESS

Stress is the measure of an external force acting over the cross sectional area of an object. Stress has units of force per area: N/m^2 (SI) or lb/in^2 (US). The SI units are commonly referred to as Pascals, abbreviated Pa. Since the 1 Pa is inconveniently small compared to the stresses most structures experience, we'll often encounter 10^3 Pa = 1 kPa (kilo Pascal), 10^6 Pa = a MPa (mega Pascal), or 10^9 Pa = GPa (giga Pascal).

There are two types of stress that a structure can experience: 1. Normal Stress and 2. Shear Stress. When a force acts perpendicular (or "normal") to the surface of an object, it exerts a normal stress. When a force acts parallel to the surface of an object, it exerts a shear stress.

Let's consider a light fixture hanging from the ceiling by a rope. The cross section of the rope is circular, and the weight of the light is pulling downward, perpendicular to the rope. This force exerts a normal stress within the rope.

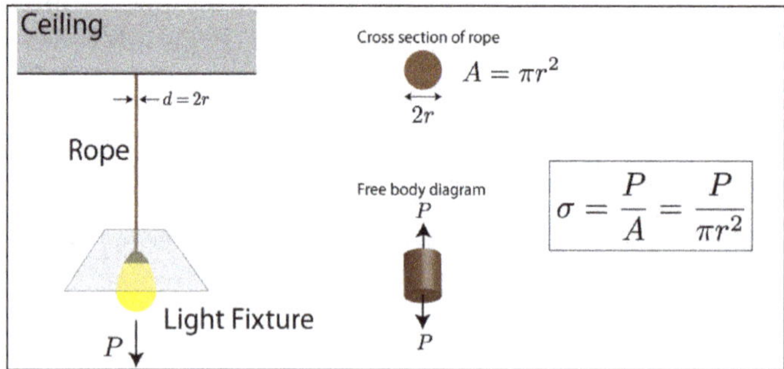

Okay, how did we arrive at this equation. There are a lot of assumptions behind the scenes. we will assume that all materials are homogenous, isotropic, and elastic. We will also assume that the object is "prismatic" – meaning the cross sections are the same all along its length (e.g. a cucumber is prismatic, while a butternut squash is not). All these assumptions allow us to state that the object will deform uniformly at every point of its cross section. The normal stress at a point on a cross section is defined as (with similar equations in the x and y directions):

$$\sigma_z = \lim_{\Delta A \to 0} \frac{\Delta F_z}{\Delta A}$$

Each small area of the cross section is subjected to the same force, and the sum of all these forces must equal the internal resultant force P. If we let ΔA go to dA, and ΔF go to dF, then we can simply integrate both sides of the equation, and we arrive at our relationship for normal stress.

$$\int dF = \int_A \sigma dA \therefore P = \sigma A$$

This relationship for the normal stress is more accurately an average normal stress, since we've averaged the internal forces over the entire cross section.

Stress is often a difficult concept to grasp because you can't easily observe it. As it turns out, placing a transparent object through cross polarized light allows you to directly observe stress within a material, based on a concept called photoelasticity.

Stress can actually exist in a material in the absence of an applied load. This is known as residual stress, and it can be used as a way to toughen materials, such as in the fabrication of the Japanese Katana sword. Conversely, undesired residual stresses can encourage crack growth and lead to fracture, such as with the collapse of the Silver Bridge of West Virginia in 1967. Perhaps the most striking example of residual stress is related to the rapid cooling of molten glass, known as "Prince Rupert's Drop".

Let's look at another example. Consider a bolt that connects two rectangular plates, and a tensile force perpendicular to the bolt. From a free body diagram, we see that the externally applied force exerts a force parallel to the circular cross section of the bolt. This external force results in a shear stress within the bolt.

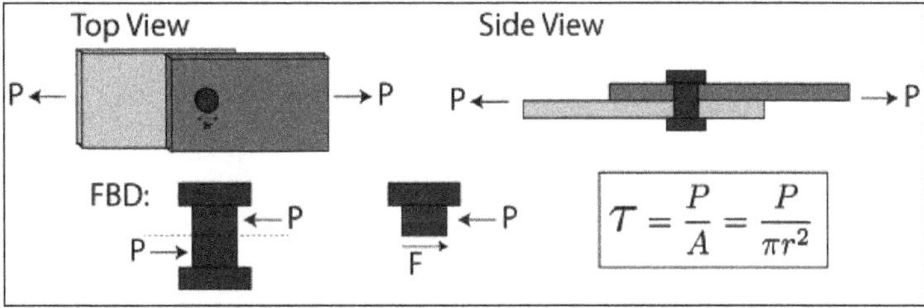

Now, the formal definitions of shear stress take on a similar form as those. Let's consider the shear stress acting on the z-face of an element:

$$T_{zx} = \lim_{\Delta A \to 0} \frac{\Delta F_x}{\Delta A}$$

$$T_{zx} = \lim_{\Delta A \to 0} \frac{\Delta F_y}{\Delta A}$$

The shear stress is the stress acting tangent to the cross section, and it takes on an average value of:

$$T_{avg} = \frac{V}{A}$$

It's important to note that the stresses we have just described are average stresses. We have assumed that all of the external force has been evenly distributed over the cross sectional area of the structure – this is not always the case, and we will revisit this assumption throughout the course.

When you look at an element under shear, things seem to be a bit more complicated. Consider a small cubic element within a structure under shear, as shown below.

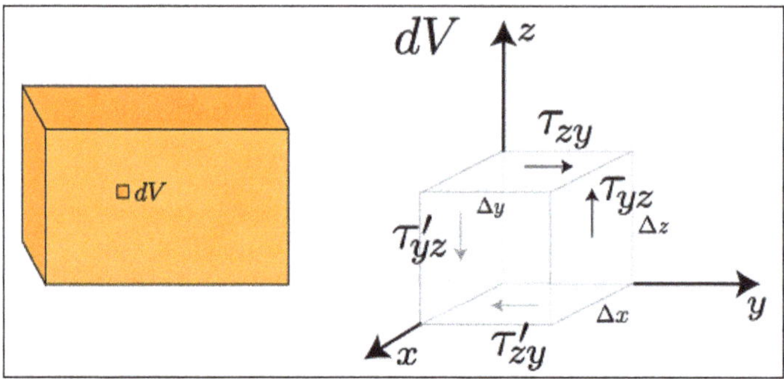

Now, equilibrium requires a shear stress acting on τ_{zy} to be accompanied by shear stresses on the other planes. But, let's consider the force equilibrium in the y-direction. Knowing that force can be written as stress (tau) times area (ΔxΔy), we can write this force equilibrium as:

$$T_{zy}(\Delta x \Delta y) - T'_{zy}(\Delta x \Delta y) = 0$$

Since the areas of the cube are by definition the same, that means $\tau_{zy} = \tau›_{zy}$. A similar force equilibrium in the z-direction leads to $\tau_{yz} = \tau›_{yz}$. Let's consider a moment equilibrium about the x-axis. Knowing that we can write the force as before, and the moment arm will be Δz, this moment balance becomes:

$$-T_{zy} = T'_{zy} = T_{yz} = T'_{yz}$$

Factor of Safety

Engineers use stress to aid in the design of structures. The external load and the geometry of the structure tells us what stress is being exerted within the material, but it tells us nothing about the material itself. Each material has an ultimate stress – a measure of how much stress the material can withstand before failing. To properly design a safe structure, we need to ensure that the applied stress from the external loading never exceeds the ultimate stress of the material. Part of the difficulty with this task is that we don't always know exactly what the external load is – it may vary unpredictably, and the structure may have to withstand unexpectedly high loads. To account for this uncertainty, we incorporate a Factor of Safety into our design. The factor of safety is just a ratio of the failure load or stress to the allowable load or stress. The failure or ultimate value is a material property while the allowable value is determined by the external force and the geometry of the structure.

$$F.S = \frac{F_{ult}}{F_{all}} \text{ or } F.S. = \frac{\sigma_{ult}}{\sigma_{all}}.$$

STRAIN

Strain, in physical sciences and engineering is the number that describes relative deformation or change in shape and size of elastic, plastic, and fluid materials under applied forces. The deformation, expressed by strain, arises throughout the material as the particles (molecules, atoms, ions) of which the material is composed are slightly displaced from their normal position.

Strains may be divided into normal strains and shear strains on the basis of the forces that cause the deformation. A normal strain is caused by forces perpendicular to planes or cross-sectional areas of the material, such as in a volume that is under pressure on all sides or in a rod that is pulled or compressed lengthwise.

A shear strain is caused by forces that are parallel to, and lie in, planes or cross-sectional areas, such as in a short metal tube that is twisted about its longitudinal axis.

In deformation of volumes under pressure, the normal strain, expressed mathematically, is equal to the change in volume divided by the original volume. In the case of elongation, or lengthwise compression, the normal strain is equal to the change in length divided by the original length. In each case the quotient of the two quantities of the same dimension is itself a pure number without dimensions. In some applications, the change (decrease) in volume or in length for compression is taken to be negative, whereas the change (increase) for dilation or tension is designated as positive. Compressive strains, by this convention, are negative, and tensile strains are positive.

In shear strain, right angles (90° angles) within the material become changed in size, as squares are deformed into diamond shapes the angles of which depart from 90°. Thus, in the illustration of the metal tube, the right angle CAF in the unstrained tube decreases to the acute angle BAF when the tube is twisted. The change in the right angle is, therefore, equal to angle BAC the tangent of which, by definition, is the ratio of $\overline{BC}$ divided by $\overline{AC}$. This ratio is the shear strain, the value of which is zero for no deformation and becomes increasingly greater as angle BAC increases. Shear strains are also dimensionless.

STRESS–STRAIN CURVE

The relationship between the stress and strain that a particular material displays is known as that particular material's stress–strain curve. It is unique for each material and is found by recording the amount of deformation (strain) at distinct intervals of a variety of loadings (stress). These curves reveal many of the properties of a material (including data to establish the Modulus of Elasticity, E).

Generally speaking, curves representing the relationship between stress and strain in any form of deformation can be regarded as stress-strain curves. The stress and strain can be normal, shear, or mixture, also can be uniaxial, biaxial, or multiaxial, even change with time. The form of deformation can be compression, stretching, torsion, rotation, and so on. If not mentioned otherwise, stress–strain curve refers to the relationship between axial normal stress and axial normal strain of materials measured in a tension test.

Consider a bar of original cross sectional area A being subjected to equal and opposite forces F pulling at the ends so the bar is under tension. The material is experiencing a stress defined to be the ratio of the force to the cross sectional area of the bar, as well as an axial elongation:

$$\sigma = \frac{F}{A_0}$$

$$\epsilon = \frac{L-L_0}{L_0} = \frac{\Delta L}{L_0}$$

Subscript 0 denotes the original dimensions of the sample. The SI unit for stress is newton per square metre, or pascal (1 pascal = 1 Pa = 1 N/m^2), and for strain is "1". Stress-strain curve for this material is plotted by elongating the sample and recording the stress variation with strain till the sample fractures. By convention, the strain is set to the horizontal axis and stress is set to vertical axis. Note that for engineering purposes we often assume the cross-section area of the material does not change during the whole deformation process. This is not true since the actual area will decrease while deforming due to elastic and plastic deformation. The curve assuming the cross-section area is fixed is called the "engineering stress-strain curve", while the curve based on the actual cross-section area is the "true stress-strain curve".

Stages

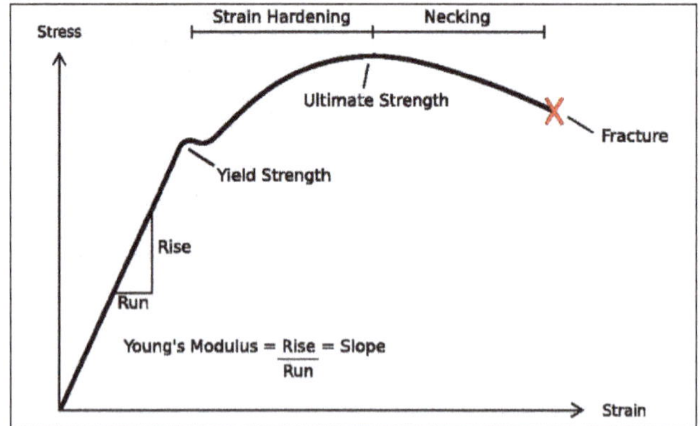

The schematic diagram for stress-strain curve of low carbon steel.

A schematic diagram for the stress-strain curve of low carbon steel at room temperature is shown in figure. There are several stages showing different behaviors, which suggests different mechanical properties. To clarify, materials can miss one or more stages shown in figure, or have totally different stages.

The first stage is the linear elastic region. The stress is proportional to the strain, that is, obeys the general Hooke's law, and the slope is Young's modulus. In this region, the material undergoes only elastic deformation. The end of the stage is the initiation point of plastic deformation. The stress component of this point is defined as yield strength (or upper yield point, UYP for short).

The second stage is the strain hardening region. This region starts as the strain goes beyond yielding point, and ends at the ultimate strength point, which is the maximal stress shown in the stress-strain curve (tensile strength, T.S., also sometimes referred to as the ultimate tensile strength, D.T.S.). In this region, the stress mainly increases as material elongates, except that there is a nearly flat region at the beginning. The stress of the flat region is defined as the lower yield point (LYP) and results from the formation and propagation of Lüders bands. Explicitly, heterogeneous plastic deformation forms bands at the upper yield strength and these bands carrying with deformation spread along the sample at the lower yield strength. After the sample is again uniformly deformed, the increase of stress with the progress of extension results from work strengthening, that is, dense dislocations induced by plastic deformation hampers the further motion of dislocations. To overcome these obstacles, a higher resolved shear stress should be applied. As the strain accumulates, work strengthening gets reinforced, till the stress reaches the tensile strength.

The third stage is the necking region. Beyond tensile strength, a *neck* forms where the local cross-sectional area becomes significantly smaller than the average. The necking deformation is heterogenous and will reinforce itself as the stress concentrates more at small section. Such positive feedback leads to quick development of necking and leads to fracture. Note that though the pulling force is decreasing, the work strengthening is still progressing, that is, the true stress keeps growing but the engineering stress decreases because the shrinking section area is not considered. This region ends up with the fracture. After fracture, percent elongation and reduction in section area can be calculated.

Relationship with True Stress and Strain

Due to the shrinking of section area and the ignored effect of developed elongation to further elongation, true stress and strain are different from engineering stress and strain.

$$\sigma_t = \frac{F}{A}$$

$$\epsilon_t = \int \frac{\delta L}{L}$$

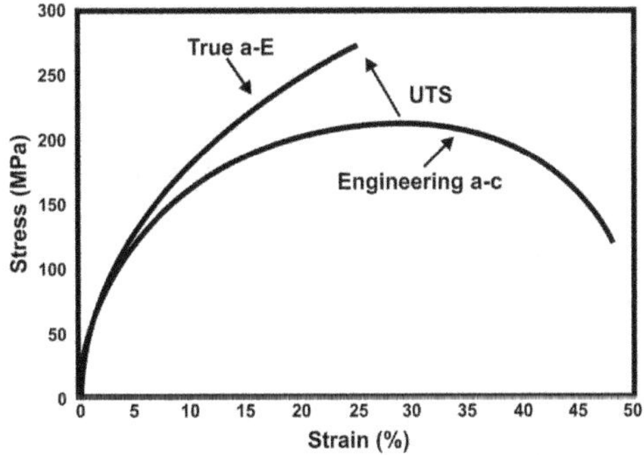

The difference between true stress-strain curve and engineering stress-strain curve.

Here the dimensions are instant values. Assuming volume of the sample conserves and deformation happens uniformly,

$$A_0 L_0 = AL$$

The true stress and strain can be expressed by engineering stress and strain. For true stress,

$$\sigma_t = \frac{F}{A} = \frac{F}{A_0} * \frac{A_0}{A} = \frac{F}{A_0} * \frac{L}{L_0} = \sigma(1+\epsilon)$$

For the strain,

$$\delta\epsilon_t = \frac{\delta L}{L}$$

Integrate both sides and apply the boundary condition,

$$\epsilon_t = ln(\frac{L}{L_0}) = ln(1+\epsilon)$$

So in a tension test, true stress is larger than engineering stress and true strain is less than engineering strain. Thus, a point defining true stress-strain curve is displaced upwards and to the left to define the equivalent engineering stress-strain curve. The difference between the true and engineering stresses and strains will increase with plastic deformation. At low strains (such as elastic deformation), the differences between the two is negligible. As for the tensile strength point, it is the maximal point in engineering stress-strain curve but is not a special point in ture stress-strain curve. Because engineering stress is proportional to the force applied along the sample, the criterion for necking formation can be set as $\delta F = 0$.

$$\delta F = 0 = \sigma_t \delta A + A \delta\sigma_t$$

$$-\frac{\delta A}{A} = \frac{\delta\sigma_t}{\sigma_t}$$

This analysis suggests nature of the UTS point. The work strengthening effect is exactly balanced by the shrinking of section area at UTS point.

After the formation of necking, the sample undergoes heterogeneous deformation, so equations above are not valid. The stress and strain at the necking can be expressed as:

$$\sigma_t = \frac{F}{A_{neck}}$$

$$\epsilon_t = ln(\frac{A_0}{A_{neck}})$$

An empirical equation is commonly used to describe the relationship between true stress and strain.

$$\sigma_t = K(\epsilon_t)^n$$

Here, n is strain-hardening coefficient and K is the strength coefficient. n is a measure of material' work hardening behavior. Materials with a higher n has a greater resistance to necking. Typically, metals at room temperature has n ranging from 0.02 to 0.5.

Classification

It is possible to distinguish some common characteristics among the stress–strain curves of various groups of materials and, on this basis, to divide materials into two broad categories; namely, the ductile materials and the brittle materials.

Ductile Materials

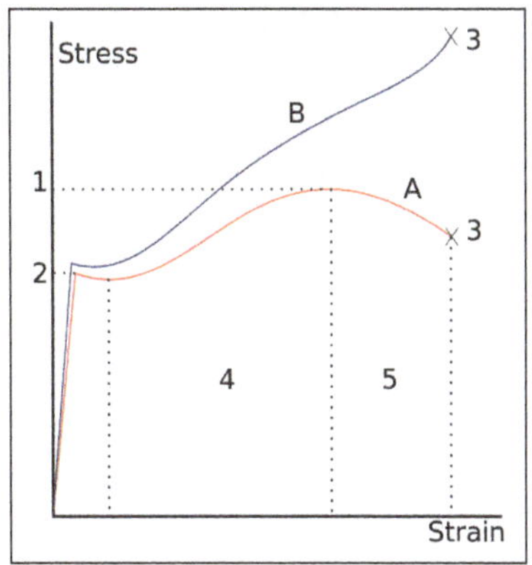

A stress–strain curve typical of structural steel.

- 1: Ultimate strength
- 2: Yield strength (yield point)
- 3: Rupture
- 4: Strain hardening region
- 5: Necking region
- A: Apparent stress (F/A_o)
- B: Actual stress (F/A)

Ductile materials, which includes structural steel and many alloys of other metals, are characterized by their ability to yield at normal temperatures.

Low carbon steel generally exhibits a very linear stress–strain relationship up to a well defined yield point. The linear portion of the curve is the elastic region and the slope is the modulus of elasticity or Young's Modulus . Many ductile materials including some metals, polymers and ceramics exhibit a yield point. Plastic flow initiates at the upper yield point and continues at the lower one. At lower yield point, permanent deformation is heterogeneously distributed along the sample. The deformation band which formed at the upper yield point will propagate along the gauge length at the lower yield point. The band occupies the whole of the gauge at the luders strain. Beyond this point, work hardening commences. The appearance of the yield point is associated with pinning of dislocations in the system. For example, solid solution interacts with dislocations and acts as pin and prevent dislocation from moving. Therefore, the stress needed to initiate the movement will be large. As long as the dislocation escape from the pinning, stress needed to continue it is less.

After the yield point, the curve typically decreases slightly because of dislocations escaping from Cottrell atmospheres. As deformation continues, the stress increases on account of strain hardening until it reaches the ultimate tensile stress. Until this point, the cross-sectional area decreases uniformly because of Poisson contractions. Then it starts necking and finally fractures.

The appearance of necking in ductile materials is associated with geometrical instability in the system. Due to the natural inhomogeneity of the material, it is common to find some regions with small inclusions or porosity within it or surface, where strain will concentrate, leading to a locally smaller area than other regions. For strain less than the ultimate tensile strain, the increase of work-hardening rate in this region will be greater than the area reduction rate, thereby make this region harder to be further deform than others, so that the instability will be removed, i.e. the materials have abilities to weaken the inhomogeneity before reaching ultimate strain. However, as the strain become larger, the work hardening rate will decreases, so that for now the region with smaller area is weaker than other region, therefore reduction in area will concentrate in this region and the neck becomes more and more pronounced until fracture. After the neck has formed in the materials, further plastic deformation is concentrated in the neck while the remainder of the material undergoes elastic contraction owing to the decrease in tensile force.

The stress-strain curve for a ductile material can be approximated using the Ramberg-Osgood equation. This equation is straightforward to implement, and only requires the material's yield strength, ultimate strength, elastic modulus, and percent elongation.

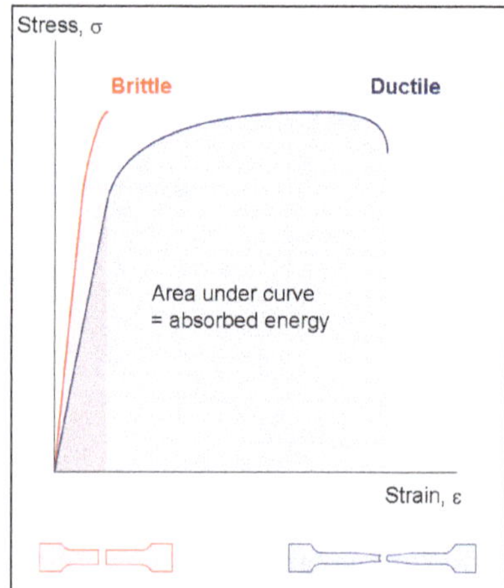

Stress–strain curve for brittle materials compared to ductile materials.

Brittle Materials

Brittle materials, which includes cast iron, glass, and stone, are characterized by the fact that rupture occurs without any noticeable prior change in the rate of elongation, sometimes they fracture before yielding.

Brittle materials such as concrete or carbon fiber do not have a well-defined yield point, and do not strain-harden. Therefore, the ultimate strength and breaking strength are the same. A typical stress–strain curve is shown in figure. Typical brittle materials like glass do not show any plastic deformation but fail while the deformation is elastic. One of the characteristics of a brittle failure is that the two broken parts can be reassembled to produce the same shape as the original component as there will not be a neck formation like in the case of ductile materials. A typical stress–strain curve for a brittle material will be linear. For some materials, such as concrete, tensile strength is negligible compared to the compressive strength and it is assumed zero for many engineering applications. Glass fibers have a tensile strength stronger than steel, but bulk glass usually does not. This is because of the stress intensity factor associated with defects in the material. As the size of the sample gets larger, the size of defects also grows. In general, the tensile strength of a rope is always less than the sum of the tensile strengths of its individual fibers.

Influencing Factors

The stress–strain curves of various materials vary widely, as a result of various intrinsic structures and compositions. Due to extrinsic factors, different tensile tests conducted on the same material yield different results, mainly depending upon the temperature of the specimen and the speed of the loading. The boundary between intrinsic and extrinsic factors is not rigid, if further mechanism is illustrated. Many factors can have an influence on the stress-strain curve by adjusting Young' modulus, strengening, toughening, as long as they modify the structure and compositions.

Time is often neglected in the stress-strain curve relations, but at higher strain rates, higher stresses will occur according to the relationship,

$$\sigma_t = K(\dot{\epsilon}_T)^m$$

where m is the strain rate sensitivity. The higher m is, the greater resistance to necking this material will have, just like the case of work-hardening coefficient.

Another dominant factor is the temperature. Temperature controls the activation of dislocations and diffusions. As the temperature increases, brittle materials can be transformed into ductile materials.

Young's Modulus

Young's modulus or Young modulus is a mechanical property that measures the stiffness of a solid material. It defines the relationship between stress (force per unit area) and strain (proportional deformation) in a material in the linear elasticity regime of a uniaxial deformation.

Young's modulus is named after the 19th-century British scientist Thomas Young. However, the concept was developed in 1727 by Leonhard Euler, and the first experiments that used the concept of Young's modulus in its current form were performed by the Italian scientist Giordano Riccati in 1782, pre-dating Young's work by 25 years.

Linear Elasticity

A solid material will undergo elastic deformation when a small load is applied to it in compression or extension. Elastic deformation is reversible (the material returns to its original shape after the load is removed).

At near-zero stress and strain, the stress–strain curve is linear, and the relationship between stress and strain is described by Hooke's law that states stress is proportional to strain. The coefficient of proportionality is Young's modulus. The higher the modulus, the more stress is needed to create the same amount of strain; an idealized rigid body would have an infinite Young's modulus.

Not many materials are linear and elastic beyond a small amount of deformation.

Formula and Units

$$E = \frac{\sigma}{\epsilon},$$

where

- E is Young's modulus
- σ is the uniaxial stress, or uniaxial force per unit surface
- ϵ is the strain, or proportional deformation (change in length divided by original length); it is dimensionless

Both E and σ have units of pressure, while ϵ is dimensionless. Young's moduli are typically so large that they are expressed not in pascals but in megapascals (MPa or N/mm^2) or gigapascals (GPa or kN/mm^2).

Material stiffness should not be confused with these properties:

- Strength: Maximal amount of stress the material can withstand while staying in the elastic (reversible) deformation regime;
- Geometric Stiffness: A global characteristic of the body that depends on its shape, and not only on the local properties of the material; for instance, an I-beam has a higher bending stiffness than a rod of the same material for a given mass per length;
- Hardness: Relative resistance of the material's surface to penetration by a harder body;
- Toughness: Amount of energy that a material can absorb before fracture.

Usage

Young's modulus enables the calculation of the change in the dimension of a bar made of an isotropic elastic material under tensile or compressive loads. For instance, it predicts how much a material sample extends under tension or shortens under compression. The Young's modulus directly applies to cases of uniaxial stress, that is tensile or compressive stress in one direction and no stress in the other directions. Young's modulus is also used in order to predict the deflection that will occur in a statically determinate beam when a load is applied at a point in between the beam's supports. Other elastic calculations usually require the use of one additional elastic property, such as the shear modulus, bulk modulus or Poisson's ratio. Any two of these parameters are sufficient to fully describe elasticity in an isotropic material.

Linear Versus Non-linear

Young's modulus represents the factor of proportionality in Hooke's law, which relates the stress and the strain. However, Hooke's law is only valid under the assumption of an *elastic* and *linear* response. Any real material will eventually fail and break when stretched over a very large distance or with a very large force; however all solid materials exhibit nearly Hookean behavior for small enough strains or stresses. If the range over which Hooke's law is valid is large enough compared to the typical stress that one expects to apply to the material, the material is said to be linear. Otherwise (if the typical stress one would apply is outside the linear range) the material is said to be non-linear.

Steel, carbon fiber and glass among others are usually considered linear materials, while other materials such as rubber and soils are non-linear. However, this is not an absolute classification: if very small stresses or strains are applied to a non-linear material, the response will be linear, but if very high stress or strain is applied to a linear material, the linear theory will not be enough. For example, as the linear theory implies reversibility, it would be absurd to use the linear theory to describe the failure of a steel bridge under a high load; although steel is a linear material for most applications, it is not in such a case of catastrophic failure.

In solid mechanics, the slope of the stress–strain curve at any point is called the tangent modulus. It can be experimentally determined from the slope of a stress–strain curve created during tensile tests conducted on a sample of the material.

Directional Materials

Young's modulus is not always the same in all orientations of a material. Most metals and ceramics, along with many other materials, are isotropic, and their mechanical properties are the same in all orientations. However, metals and ceramics can be treated with certain impurities, and metals can be mechanically worked to make their grain structures directional. These materials then become anisotropic, and Young's modulus will change depending on the direction of the force vector. Anisotropy can be seen in many composites as well. For example, carbon fiber has a much higher Young's modulus (is much stiffer) when force is loaded parallel to the fibers (along the grain). Other such materials include wood and reinforced concrete. Engineers can use this directional phenomenon to their advantage in creating structures.

Calculation

Young's modulus E, can be calculated by dividing the tensile stress, $\sigma(\varepsilon)$, by the engineering extensional strain, ε, in the elastic (initial, linear) portion of the physical stress–strain curve:

$$E \equiv \frac{\sigma(\varepsilon)}{\varepsilon} = \frac{F / A}{\Delta L / L_0} = \frac{F L_0}{A \Delta L}$$

where,

E is the Young's modulus (modulus of elasticity)

F is the force exerted on an object under tension;

A is the actual cross-sectional area, which equals the area of the cross-section perpendicular to the applied force;

ΔL is the amount by which the length of the object changes (ΔL is positive if the material is stretched , and negative when the material is compressed);

L_o is the original length of the object.

Force Exerted by Stretched or Contracted Material

The Young's modulus of a material can be used to calculate the force it exerts under specific strain.

$$F = \frac{EA\Delta L}{L_0}$$

where F is the force exerted by the material when contracted or stretched by ΔL.

Hooke's law for a stretched wire can be derived from this formula:

$$F = \left(\frac{EA}{L_0}\right)\Delta L = kx$$

where it comes in saturation

$$k \equiv \frac{EA}{L_0} \text{ and } x \equiv \Delta L.$$

But note that the elasticity of coiled springs comes from shear modulus, not Young's modulus.

Elastic Potential Energy

The elastic potential energy stored in a linear elastic material is given by the integral of the Hooke's law:

$$U_e = \int kx\,dx = \frac{1}{2}kx^2.$$

now by explicating the intensive variables:

$$U_e = \int \frac{EA\Delta L}{L_0}\,d\Delta L = \frac{EA}{L_0}\int \Delta L\,d\Delta L = \frac{EA\Delta L^2}{2L_0}$$

This means that the elastic potential energy density (i.e., per unit volume) is given by:

$$\frac{U_e}{AL_0} = \frac{E\Delta L^2}{2L_0^2}$$

or, in simple notation, for a linear elastic material: $u_e(\varepsilon) = \int E\,\varepsilon\,d\varepsilon = \frac{1}{2}E\varepsilon^2$, since the strain is defined $\varepsilon \equiv \frac{\Delta L}{L_0}$.

In a nonlinear elastic material the Young's modulus is a function of the strain, so the second equivalence no longer holds and the elastic energy is not a quadratic function of the strain.

$$u_e(\varepsilon) = \int E(\varepsilon)\,\varepsilon\, d\varepsilon \neq \frac{1}{2}E\varepsilon^2$$

Relation among Elastic Constants

For homogeneous isotropic materials simple relations exist between elastic constants (Young's modulus E, shear modulus G, bulk modulus K, and Poisson's ratio ν) that allow calculating them all as long as two are known:

$$E = 2G(1+\nu) = 3K(1-2\nu)$$

Temperature Dependence

The Young's modulus of metals varies with the temperature and can be realized through the change in the interatomic bonding of the atoms and hence its change is found to be dependent on the change in the work function of the metal. Although classically, this change is predicted through fitting and without a clear underlying mechanism (e.g. the Watchman's formula), the Rahemi-Li model demonstrates how the change in the electron work function leads to change in the Young's modulus of metals and predicts this variation with calculable parameters, using the generalization of the Lennard-Jones potential to solids. In general, as the temperature increases, the Young's modulus decreases via $E(T) = \beta(\phi(T))^6$ Where the electron work function varies with the temperature as $\phi(T) = \phi_0 - \gamma \frac{(k_B T)^2}{\phi_0}$ and γ is a calculable material property which is dependent on the crystal structure (e.g. BCC, FCC, etc.) ϕ_0 is the electron work function at T=0 and β is constant throughout the change.

Approximate Values

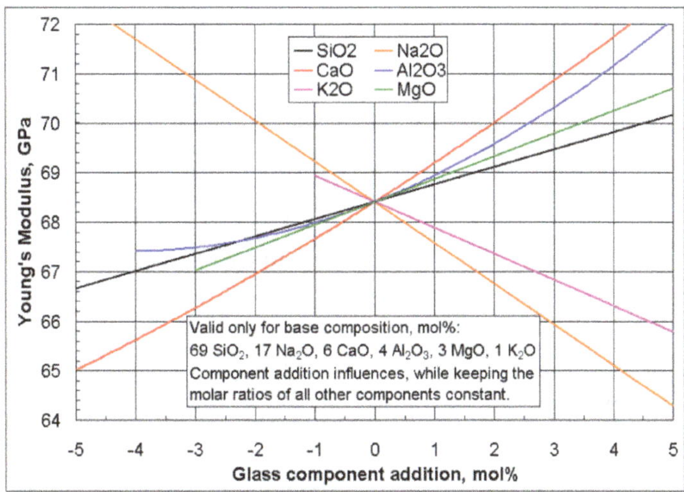

Influences of selected glass component additions on Young's modulus of a specific base glass.

Young's modulus can vary somewhat due to differences in sample composition and test method. The rate of deformation has the greatest impact on the data collected, especially in polymers.

YIELD STRENGTH

The yield point is the point on a *stress–strain curve* that indicates the limit of elastic behavior and the beginning plastic behavior. Yield strength or yield stress is the material property defined as the stress at which a material begins to deform plastically whereas yield point is the point where nonlinear (elastic + plastic) deformation begins. Prior to the yield point the material will deform elastically and will return to its original shape when the applied stress is removed. Once the yield point is passed, some fraction of the deformation will be permanent and non-reversible.

The yield point determines the limits of performance for mechanical components, since it represents the upper limit to forces that can be applied without permanent deformation. In structural engineering, this is a soft failure mode which does not normally cause catastrophic failure or ultimate failure unless it accelerates buckling.

Advances in measurement techniques allow higher precision mapping of the yield point which, as Marcus Reiner stated, showed "there was no yield point"

Yield strength is the critical material property exploited by many fundamental techniques of material-working: to reshape material with pressure (such as forging, rolling, pressing, bending, extruding, or hydroforming), to separate material by cutting (such as machining) or shearing, and to join components rigidly with fasteners. Yield load can be taken as the load applied to the centre of a carriage spring to straighten its leaves.

The offset yield point (or proof stress) is the stress at which 0.2% plastic deformation occurs.

In the three-dimensional principal stresses ($\sigma_1, \sigma_2, \sigma_3$), an infinite number of yield points form together a yield surface.

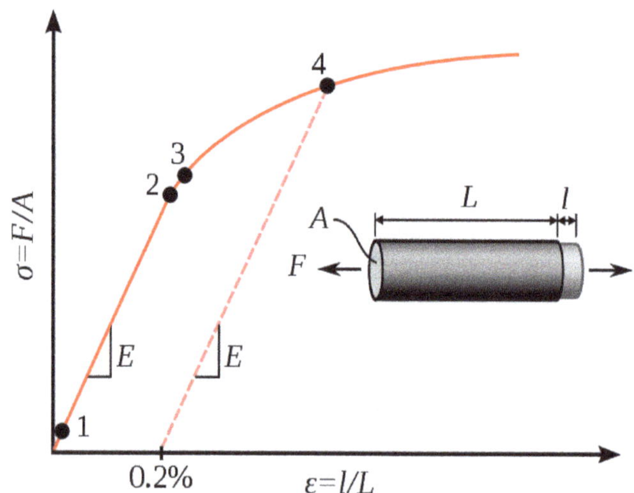

Stress–strain curve showing typical yield behavior for nonferrous alloys. (Stress, σ, shown as a function of strain, ϵ.)

1. True elastic limit
2. Proportionality limit
3. Elastic limit
4. Offset yield strength

It is often difficult to precisely define yielding due to the wide variety of stress–strain curves exhibited by real materials. In addition, there are several possible ways to define yielding:

True Elastic Limit

The lowest stress at which dislocations move. This definition is rarely used, since dislocations move at very low stresses, and detecting such movement is very difficult.

Proportionality Limit

Up to this amount of stress, stress is proportional to strain (Hooke's law), so the stress–strain graph is a straight line, and the gradient will be equal to the elastic modulus of the material.

Elastic Limit (Yield Strength)

Beyond the elastic limit, permanent deformation will occur. The elastic limit is therefore the lowest stress point at which permanent deformation can be measured. This requires a manual load-unload procedure, and the accuracy is critically dependent on the equipment used and operator skill. For elastomers, such as rubber, the elastic limit is much larger than the proportionality limit. Also, precise strain measurements have shown that plastic strain begins at low stresses.

Yield Point

The point in the stress–strain curve at which the curve levels off and plastic deformation begins to occur.

Offset Yield Point (Proof Stress)

When a yield point is not easily defined based on the shape of the stress–strain curve an *offset yield point* is arbitrarily defined. The value for this is commonly set at 0.1% or 0.2% plastic strain. The offset value is given as a subscript, e.g., $R_{p0.2}$ = 310 MPa or $R_{p1.0}$= 350 N/mm^2. For most practical engineering uses, this $R_{p0.2}$ is multiplied by a factor of safety to obtain a lower value of the offset yield point ($R_{v0.2}$). High strength steel and aluminum alloys do not exhibit a yield point, so this offset yield point is used on these materials.

Upper and Lower Yield Points

Some metals, such as mild steel, reach an upper yield point before dropping rapidly to a lower yield point. The material response is linear up until the upper yield point, but the lower yield point is used in structural engineering as a conservative value. If a metal is only stressed to the upper yield point, and beyond, Lüders bands can develop.

Theoretical Yield Strength

The theoretical yield strength can be estimated by considering the process of yield at the atomic level. In a perfect crystal, shearing results in the displacement of an entire plane of atoms by one interatomic separation distance, b, relative to the plane below. In order for the atoms to move, considerable force must be applied to overcome the lattice energy and move the atoms in the top plane over the lower atoms and into a new lattice site. The applied stress to overcome the resistance of a perfect lattice to shear is the theoretical yield strength, τ_{max}.

Derivation

The stress displacement curve of a plane of atoms varies sinusoidally as stress peaks when an atom is forced over the atom below and then falls as the atom slides into the next lattice point.

$$\tau = \tau_{max} \sin\left(\frac{2\pi x}{b}\right)$$

where b is the interatomic separation distance. Since $\tau = G\gamma$ and $d\tau/d\gamma = G$ at small strains (ie. Single atomic distance displacements), this equation becomes:

$$G = \frac{d\tau}{dx} = \frac{2\pi}{b}\tau_{max} \cos\left(\frac{2\pi x}{b}\right) = \frac{2\pi}{b}\tau_{max}$$

For small displacement of $\gamma = x/a$, where a is the spacing of atoms on the slip plane, this can be rewritten as:

$$G = \frac{d\tau}{d\gamma} = \frac{2\pi}{b}\tau_{max}$$

Giving a value of τ_{max} equal to:

$$\tau_{max} = \frac{Gb}{2\pi a}$$

The theoretical yield strength can be approximated as $\tau_{max} = G/30$.

Typical Values of Theoretical and Experimental Yield Stress

The theoretical yield strength of a perfect crystal is much higher than the observed stress at initiation of plastic flow. Theoretical and experimental yield stresses of common materials are shown in the table below:

Material	Theoretical Shear Strength (GPa)	Experimental Shear Strength (MPa)
Ag	1.0	0.37
Al	0.9	0.78
Cu	1.4	0.49
Ni	2.6	3.2
α-Fe	2.6	27.5

That experimentally measured yield strength is significantly lower than the expected theoretical value can be explained by the presence of dislocations and defects in the materials. Indeed, whiskers with perfect single crystal structure and defect free surfaces have been shown to demonstrate yield stress approaching the theoretical value. For example, nanowhiskers of copper were shown to undergo brittle fracture at 1 GPa, a value much higher than the strength of bulk copper and approaching the theoretical value.

Yield Criterion

A yield criterion, often expressed as yield surface, or yield locus, is a hypothesis concerning the limit of elasticity under any combination of stresses. There are two interpretations of yield criterion: one is purely mathematical in taking a statistical approach while other models attempt to provide a justification based on established physical principles. Since stress and strain are tensor qualities they can be described on the basis of three principal directions, in the case of stress these are denoted by σ_1, σ_2, and σ_3.

The following represent the most common yield criterion as applied to an isotropic material (uniform properties in all directions). Other equations have been proposed or are used in specialist situations.

Isotropic Yield Criteria

Maximum Principal Stress Theory – by W.J.M Rankine. Yield occurs when the largest principal stress exceeds the uniaxial tensile yield strength. Although this criterion allows for a quick and easy comparison with experimental data it is rarely suitable for design purposes. This theory gives good predictions for brittle materials.

$$\sigma \leq \sigma$$

Maximum Principal Strain Theory – by St.Venant. Yield occurs when the maximum principal strain reaches the strain corresponding to the yield point during a simple tensile test. In terms of the principal stresses this is determined by the equation:

$$\sigma_1 - \nu(\sigma_2 + \sigma_3) \leq \sigma_y.$$

Maximum Shear Stress Theory – Also known as the Tresca yield criterion, after the French scientist Henri Tresca. This assumes that yield occurs when the shear stress τ exceeds the shear yield strength τ_y:

$$\tau = \frac{\sigma_1 - \sigma_3}{2} \leq \tau_y.$$

Total Strain Energy Theory – This theory assumes that the stored energy associated with elastic deformation at the point of yield is independent of the specific stress tensor. Thus yield occurs when the strain energy per unit volume is greater than the strain energy at the elastic limit in simple tension. For a 3-dimensional stress state this is given by:

$$\sigma_1^2 + \sigma_2^2 + \sigma_3^2 - 2\nu(\sigma_1\sigma_2 + \sigma_2\sigma_3 + \sigma_1\sigma_3) \leq \sigma_y^2.$$

Maximum Distortion Energy Theory (von Mises yield criterion) – This theory proposes that the total strain energy can be separated into two components: the *volumetric* (hydrostatic) strain energy and the *shape* (distortion or shear) strain energy. It is proposed that yield occurs when the distortion component exceeds that at the yield point for a simple tensile test. This theory is also known as the von Mises yield criterion.

Based on a different theoretical underpinning this expression is also referred to as octahedral shear stress theory.

Other commonly used isotropic yield criteria are the

- Von Mises yield criterion
- Mohr-Coulomb yield criterion
- Drucker-Prager yield criterion
- Bresler-Pister yield criterion
- Willam-Warnke yield criterion

The yield surfaces corresponding to these criteria have a range of forms. However, most isotropic yield criteria correspond to convex yield surfaces.

Anisotropic Yield Criteria

When a metal is subjected to large plastic deformations the grain sizes and orientations change in the direction of deformation. As a result, the plastic yield behavior of the material shows directional dependency. Under such circumstances, the isotropic yield criteria such as the von Mises yield criterion are unable to predict the yield behavior accurately. Several anisotropic yield criteria have been developed to deal with such situations. Some of the more popular anisotropic yield criteria are:

- Hill's quadratic yield criterion
- Generalized Hill yield criterion
- Hosford yield criterion

Factors Influencing Yield Strength

The stress at which yield occurs is dependent on both the rate of deformation (strain rate) and, more significantly, the temperature at which the deformation occurs. In general, the yield strength increases with strain rate and decreases with temperature. When the latter is not the case, the material is said to exhibit yield strength anomaly, which is typical for superalloys and leads to their use in applications requiring high strength at high temperatures.

Early work by Alder and Philips found that the relationship between yield strength and strain rate (at constant temperature) was best described by a power law relationship of the form,

$$\sigma_y = C(\dot{\epsilon})^m$$

where C is a constant and m is the strain rate sensitivity. The latter generally increases with temperature, and materials where m reaches a value greater than ~0.5 tend to exhibit super plastic behavior. m can be found from a log-log plot of yield strength at a fixed plastic strain versus the strain rate.

$$m = \frac{\partial \ln \sigma(\epsilon)}{\partial \ln(\dot{\epsilon})}$$

Later, more complex equations were proposed that simultaneously dealt with both temperature and strain rate:

$$\sigma_y = \frac{1}{\alpha} \sinh^{-1} \left[\frac{Z}{A} \right]^{\frac{1}{n}}$$

where α and A are constants and Z is the temperature-compensated strain-rate – often described by the Zener-Hollomon parameter:

$$Z = (\dot{\epsilon}) \exp\left(\frac{Q_{HW}}{RT} \right)$$

where Q_{HW} is the activation energy for hot deformation and T is the absolute temperature.

Strengthening Mechanisms

There are several ways in which crystalline and amorphous materials can be engineered to increase their yield strength. By altering dislocation density, impurity levels, grain size (in crystalline materials), the yield strength of the material can be fine tuned. This occurs typically by introducing defects such as impurities dislocations in the material. To move this defect (plastically deforming or yielding the material), a larger stress must be applied. This thus causes a higher yield stress in the material. While many material properties depend only on the composition of the bulk material, yield strength is extremely sensitive to the materials processing as well.

These mechanisms for crystalline materials include

- Work hardening
- Solid solution strengthening
- Precipitation strengthening
- Grain boundary strengthening

Work Hardening

Where deforming the material will introduce dislocations, which increases their density in the material. This increases the yield strength of the material, since now more stress must be applied to move these dislocations through a crystal lattice. Dislocations can also interact with each other, becoming entangled.

The governing formula for this mechanism is:

$$\Delta\sigma_y = Gb\sqrt{\rho}$$

where σ_y is the yield stress, G is the shear elastic modulus, b is the magnitude of the Burgers vector, and ρ is the dislocation density.

Solid Solution Strengthening

By alloying the material, impurity atoms in low concentrations will occupy a lattice position directly below a dislocation, such as directly below an extra half plane defect. This relieves a tensile strain directly below the dislocation by filling that empty lattice space with the impurity atom.

The relationship of this mechanism goes as:

$$\Delta\tau = Gb\sqrt{C_s}\epsilon^{\frac{3}{2}}$$

where τ is the shear stress, related to the yield stress, G and b are the same as in the above example,   is the concentration of solute and ϵ is the strain induced in the lattice due to adding the impurity.

Particle/precipitate Strengthening

Where the presence of a secondary phase will increase yield strength by blocking the motion of dislocations within the crystal. A line defect that, while moving through the matrix, will be forced against a small particle or precipitate of the material. Dislocations can move through this particle either by shearing the particle, or by a process known as bowing or ringing, in which a new ring of dislocations is created around the particle.

The shearing formula goes as:

$$\Delta\tau = \frac{r_{\text{particle}}}{l_{\text{interparticle}}}\gamma_{\text{particle-matrix}}$$

and the bowing/ringing formula:

$$\Delta\tau = \frac{Gb}{l_{\text{interparticle}} - 2r_{\text{particle}}}$$

In these formulas, r_{particle} is the particle radius, $\gamma_{\text{particle-matrix}}$ is the surface tension between the matrix and the particle, $l_{\text{interparticle}}$ is the distance between the particles.

Grain Boundary Strengthening

Where a buildup of dislocations at a grain boundary causes a repulsive force between dislocations. As grain size decreases, the surface area to volume ratio of the grain increases, allowing more buildup of dislocations at the grain edge. Since it requires a lot of energy to move dislocations to another grain, these dislocations build up along the boundary, and increase the yield stress of the material. Also known as Hall-Petch strengthening, this type of strengthening is governed by the formula:

$$\sigma_y = \sigma_0 + kd^{-\frac{1}{2}}$$

where,

σ_0 is the stress required to move dislocations,

k is a material constant, and

d is the grain size.

Testing

Yield strength testing involves taking a small sample with a fixed cross-section area, and then pulling it with a controlled, gradually increasing force until the sample changes shape or breaks. This is called a Tensile Test. Longitudinal and/or transverse strain is recorded using mechanical or optical extensometers.

Yield behaviour can also be simulated using virtual tests (on computer models of materials), particularly where macroscopic yield is governed by the microstructural architecture of the material being studied.

Indentation hardness correlates roughly linearly with tensile strength for most steels, but measurements on one material cannot be used as a scale to measure strengths on another. Hardness testing can therefore be an economical substitute for tensile testing, as well as providing local variations in yield strength due to, e.g., welding or forming operations. However, for critical situations tension testing is done to eliminate ambiguity.

Implications for Structural Engineering

Yielded structures have a lower stiffness, leading to increased deflections and decreased buckling strength. The structure will be permanently deformed when the load is removed, and may have residual stresses. Engineering metals display strain hardening, which implies that the yield stress is increased after unloading from a yield state. Highly optimized structures, such as airplane beams and components, rely on yielding as a fail-safe failure mode. No safety factor is therefore needed when comparing limit loads (the highest loads expected during normal operation) to yield criteria.

Typical Yield and Ultimate Strengths

Many of the values depend on manufacturing process and purity/composition.

Material	Yield strength (MPa)	Ultimate strength (MPa)	Density (g/cm³)	Free breaking length (km)
ASTM A36 steel	250	400	7.87	3.2
Steel, API 5L X65	448	531	7.85	5.8
Steel, high strength alloy ASTM A514	690	760	7.85	9.0
Steel, prestressing strands	1650	1860	7.85	21.6
Piano wire		1740-3300	7.8	28.7
Carbon fiber (CF, CFK)		5650	1.75	329
High-density polyethylene (HDPE)	26–33	37	0.95	2.8
Polypropylene	12–43	19.7–80	0.91	1.3

Stainless steel AISI 302 – cold-rolled	520	860		
Cast iron 4.5% C, ASTM A-48		172	7.20	2.4
Titanium alloy (6% Al, 4% V)	830	900	4.51	18.8
Aluminium alloy 2014-T6	400	455	2.7	15.1
Copper 99.9% Cu	70	220	8.92	0.8
Cupronickel 10% Ni, 1.6% Fe, 1% Mn, balance Cu	130	350	8.94	1.4
Brass	Approx. 200+	550	8.5	3.8
Spider silk	1150 (??)	1400	1.31	109
Silkworm silk	500			25
Aramid (Kevlar or Twaron)	3620	3757	1.44	256.3
UHMWPE	20	35	0.97	2.1
Bone (limb)	104–121	130		3
Nylon, type 6/6	45	75		2

Elements in the annealed state			
Element	Young's modulus (GPa)	Proof or yield stress (MPa)	Ultimate tensile Strength (MPa)
Aluminium	70	15–20	40–50
Copper	130	33	210
Iron	211	80–100	350
Nickel	170	14–35	140–195
Silicon	107	5000–9000	
Tantalum	186	180	200
Tin	47	9–14	15–200
Titanium	120	100–225	240–370
Tungsten	411	550	550–620

COMPRESSIVE STRENGTH

Compressive strength or compression strength is the capacity of a material or structure to withstand loads tending to reduce size, as opposed to tensile strength, which withstands loads tending to elongate. In other words, compressive strength resists compression (being pushed together), whereas tensile strength resists tension (being pulled apart). In the study of strength of materials, tensile strength, compressive strength, and shear strength can be analyzed independently.

Some materials fracture at their compressive strength limit; others deform irreversibly, so a given amount of deformation may be considered as the limit for compressive load. Compressive strength is a key value for design of structures.

Measuring the compressive strength of a steel drum.

Compressive strength is often measured on a universal testing machine; these range from very small table-top systems to ones with over 53 MN capacity. Measurements of compressive strength are affected by the specific test method and conditions of measurement. Compressive strengths are usually reported in relationship to a specific technical standard.

When a specimen of material is loaded in such a way that it extends it is said to be in *tension*. On the other hand, if the material compresses and shortens it is said to be in *compression*.

On an atomic level, the molecules or atoms are forced apart when in tension whereas in compression they are forced together. Since atoms in solids always try to find an equilibrium position, and distance between other atoms, forces arise throughout the entire material which oppose both tension or compression. The phenomena prevailing on an atomic level are therefore similar.

The "strain" is the relative change in length under applied stress; positive strain characterises an object under tension load which tends to lengthen it, and a compressive stress that shortens an object gives negative strain. Tension tends to pull small sideways deflections back into alignment, while compression tends to amplify such deflection into buckling.

Compressive strength is measured on materials, components, and structures. By definition, the ultimate compressive strength of a material is that value of uniaxial compressive stress reached when the material fails completely. The compressive strength is usually obtained experimentally by means of a *compressive test*. The apparatus used for this experiment is the same as that used in a tensile test. However, rather than applying a uniaxial tensile load, a uniaxial compressive load is applied. As can be imagined, the specimen (usually cylindrical) is shortened as well as spread laterally. A stress–strain curve is plotted by the instrument and would look similar to the following:

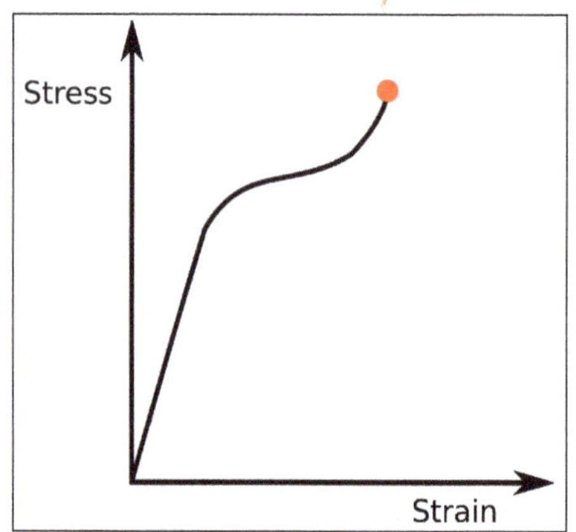

True Stress-Strain curve for a typical specimen.

The compressive strength of the material would correspond to the stress at the red point shown on the curve. In a compression test, there is a linear region where the material follows Hooke's law. Hence, for this region, $\sigma = E\epsilon$, where, this time, E refers to the Young's Modulus for compression. In this region, the material deforms elastically and returns to its original length when the stress is removed.

This linear region terminates at what is known as the yield point. Above this point the material behaves plastically and will not return to its original length once the load is removed.

There is a difference between the engineering stress and the true stress. By its basic definition the uniaxial stress is given by:

$$\sigma = \frac{F}{A}$$

where, F = Load applied [N], A = Area [m^2]

As stated, the area of the specimen varies on compression. In reality therefore the area is some function of the applied load i.e. A = f(F). Indeed, stress is defined as the force divided by the area at the start of the experiment. This is known as the engineering stress and is defined by,

$$\sigma_e = \frac{F}{A_0}$$

A_0=Original specimen area [m^2]

Correspondingly, the engineering strain would be defined by:

$$\epsilon_e = \frac{l - l_0}{l_0}$$

where, l = current specimen length [m] and l_0 = original specimen length [m]

The compressive strength would therefore correspond to the point on the engineering stress strain curve $(\epsilon_e^*, \sigma_e^*)$ defined by

$$\sigma_e^* = \frac{F^*}{A_0}$$

$$\epsilon_e^* = \frac{l^* - l_0}{l_0}$$

where, F^* = load applied just before crushing and l^* = specimen length just before crushing.

Deviation of Engineering Stress From True Stress

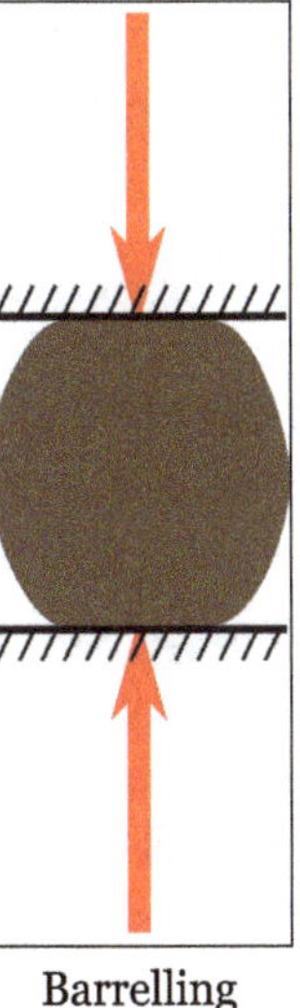

Barrelling

In engineering design practice, professionals mostly rely on the engineering stress. In reality, the *true stress* is different from the engineering stress. Hence calculating the compressive strength of a material from the given equations will not yield an accurate result. This is because the cross sectional area A_0 changes and is some function of load $A = \varphi(F)$.

The difference in values may therefore be summarized as follows:

- On compression, the specimen will shorten. The material will tend to spread in the lateral direction and hence increase the cross sectional area.

- In a compression test the specimen is clamped at the edges. For this reason, a frictional force arises which will oppose the lateral spread. This means that work has to be done to oppose this frictional force hence increasing the energy consumed during the process. This results in a slightly inaccurate value of stress obtained from the experiment. The frictional force is not constant for the entire cross section of the specimen. It varies from a minimum at the centre, away from the clamps, to a maximum at the edges where it is clamped. Due to this, a phenomenon known as *barrelling* occurs where the specimen attains a barrel shape.

Comparison of Compressive and Tensile Strengths

Concrete and ceramics typically have much higher compressive strengths than tensile strengths. Composite materials, such as glass fiber epoxy matrix composite, tend to have higher tensile strengths than compressive strengths. Metals are difficult to test to failure in tension vs compression. In compression metals fail from buckling/crumbling/45deg shear which is much different (though higher stresses) than tension which fails from defects or necking down.

Typical Values

Material	R_s [MPa]
Porcelain	500
Bone	150
concrete	20-80
Ice (0 °C)	3
Styrofoam	~1

Compressive Strength of Concrete

For designers, compressive strength is one of the most important engineering properties of concrete. It is a standard industrial practice that the concrete is classified based on grades. This grade is nothing but the Compressive Strength of the concrete cube or cylinder. Cube or Cylinder samples are usually tested under a compression testing machine to obtain the compressive strength of concrete. The test requisites differ country to country based on the design code. As per Indian codes, compressive strength of concrete is defined as the compressive strength of concrete is given in terms of the characteristic compressive strength of 150 mm size cubes tested at 28 days (fck). The characteristic strength is defined as the strength of the concrete below which not more than 5% of the test results are expected to fall.

For design purposes, this compressive strength value is restricted by dividing with a factor of safety, whose value depends on the design philosophy used.

ULTIMATE TENSILE STRENGTH

Ultimate tensile strength (UTS), often shortened to tensile strength (TS), ultimate strength, or Ftu within equations, is the capacity of a material or structure to withstand loads tending to elongate, as opposed to compressive strength, which withstands loads tending to reduce size. In other words, tensile strength resists tension (being pulled apart), whereas compressive strength resists compression (being pushed together). Ultimate tensile strength is measured by the maximum stress that a material can withstand while being stretched or pulled before breaking. In the study of strength of materials, tensile strength, compressive strength, and shear strength can be analyzed independently.

Some materials break very sharply, without plastic deformation, in what is called a brittle failure. Others, which are more ductile, including most metals, experience some plastic deformation and possibly necking before fracture.

The UTS is usually found by performing a tensile test and recording the engineering stress versus strain. The highest point of the stress–strain curve is the UTS. It is an intensive property; therefore its value does not depend on the size of the test specimen. However, it is dependent on other factors, such as the preparation of the specimen, the presence or otherwise of surface defects, and the temperature of the test environment and material.

Tensile strengths are rarely used in the design of ductile members, but they are important in brittle members. They are tabulated for common materials such as alloys, composite materials, ceramics, plastics, and wood.

Tensile strength can be defined for liquids as well as solids under certain conditions. For example, when a tree draws water from its roots to its upper leaves by transpiration, the column of water is pulled upwards from the top by the cohesion of the water in the xylem, and this force is transmitted down the column by its tensile strength. Air pressure, osmotic pressure, and capillary tension also plays a small part in a tree's ability to draw up water, but this alone would only be sufficient to push the column of water to a height of less than ten metres, and trees can grow much higher than that (over 100 m).

Tensile strength is defined as a stress, which is measured as force per unit area. For some non-homogeneous materials (or for assembled components) it can be reported just as a force or as a force per unit width. In the International System of Units (SI), the unit is the pascal (Pa) (or a multiple thereof, often megapascals (MPa), using the SI prefix *mega*); or, equivalently to pascals, newtons per square metre (N/m^2). A United States customary unit is pounds per square inch (lb/in^2 or psi), or kilo-pounds per square inch (ksi, or sometimes kpsi), which is equal to 1000 psi; kilo-pounds per square inch are commonly used in one country (US), when measuring tensile strengths.

Ductile Materials

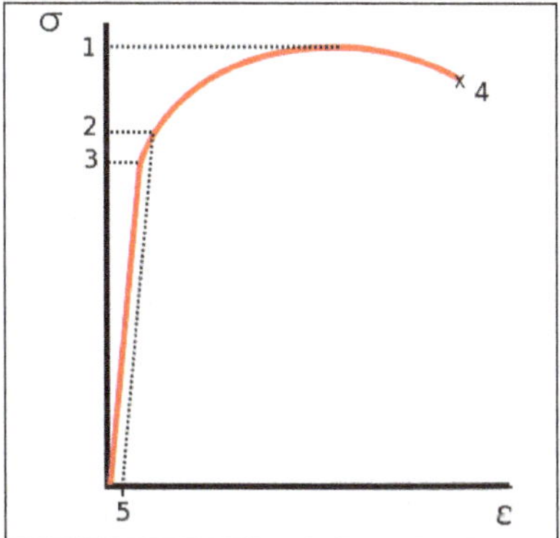

"Engineering" stress–strain (σ–ε) curve typical of aluminum
1. Ultimate strength 2. Yield strength 3. Proportional limit stress 4. Fracture 5. Offset strain (typically 0.2%)

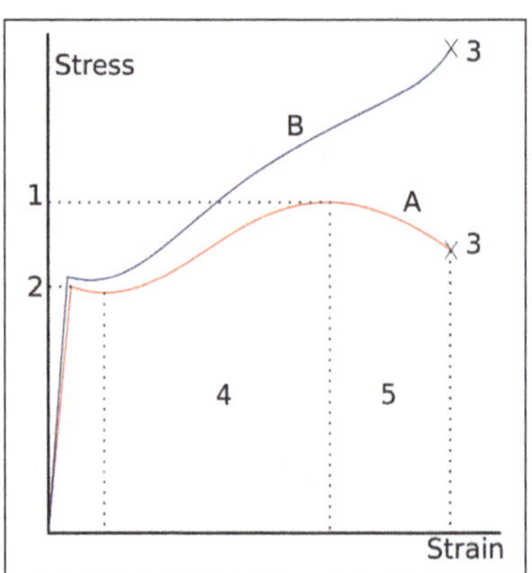

"Engineering" (red) and "true" (blue) stress–strain curve typical of structural steel.
1: Ultimate strength 2: Yield strength (yield point) 3: Rupture 4: Strain hardening region 5: Necking region
A: Apparent stress (F/A_0) B: Actual stress (F/A).

Many materials can display linear elastic behavior, defined by a linear stress–strain relationship, as shown in figure up to point 3. The elastic behavior of materials often extends into a non-linear region, represented in figure by point 2 (the "yield point"), up to which deformations are completely recoverable upon removal of the load; that is, a specimen loaded elastically in tension will elongate, but will return to its original shape and size when unloaded. Beyond this elastic region, for ductile materials, such as steel, deformations are plastic. A plastically deformed specimen does not completely return to its original size and shape when unloaded. For many applications, plastic deformation is unacceptable, and is used as the design limitation.

After the yield point, ductile metals undergo a period of strain hardening, in which the stress increases again with increasing strain, and they begin to neck, as the cross-sectional area of the specimen decreases due to plastic flow. In a sufficiently ductile material, when necking becomes substantial, it causes a reversal of the engineering stress–strain curve; this is because the *engineering stress* is calculated assuming the original cross-sectional area before necking. The reversal point is the maximum stress on the engineering stress–strain curve, and the engineering stress coordinate of this point is the ultimate tensile strength, given by point 1.

UTS is not used in the design of ductile static members because design practices dictate the use of the yield stress. It is, however, used for quality control, because of the ease of testing. It is also used to roughly determine material types for unknown samples.

The UTS is a common engineering parameter to design members made of brittle material because such materials have no yield point.

Testing

Typically, the testing involves taking a small sample with a fixed cross-sectional area, and then pulling it with a tensometer at a constant strain (change in gauge length divided by initial gauge length) rate until the sample breaks.

When testing some metals, indentation hardness correlates linearly with tensile strength. This important relation permits economically important nondestructive testing of bulk metal deliveries with lightweight, even portable equipment, such as hand-held Rockwell hardness testers. This practical correlation helps quality assurance in metalworking industries to extend well beyond the laboratory and universal testing machines.

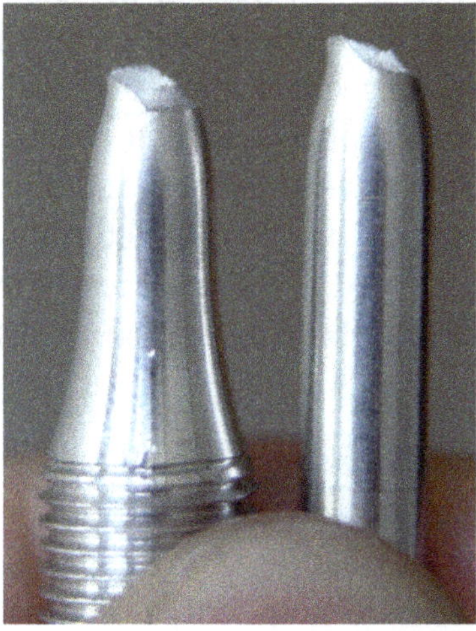

Round bar specimen after tensile stress testing.

HARDNESS

Hardness is the resistance of a material to localized deformation. The term can apply to deformation from indentation, scratching, cutting or bending. In metals, ceramics and most polymers, the deformation considered is plastic deformation of the surface. For elastomers and some polymers, hardness is defined at the resistance to elastic deformation of the surface. The lack of a fundamental definition indicates that hardness is not be a basic property of a material, but rather a composite one with contributions from the yield strength, work hardening, true tensile strength, modulus, and others factors. Hardness measurements are widely used for the quality control of materials because they are quick and considered to be nondestructive tests when the marks or indentations produced by the test are in low stress areas.

There are a large variety of methods used for determining the hardness of a substance. A few of the more common methods are introduced below.

Mohs Hardness Test

One of the oldest ways of measuring hardness was devised by the German mineralogist Friedrich Mohs in 1812. The Mohs hardness test involves observing whether a materials surface is scratched by a substance of known or defined hardness. To give numerical values to this physical property, minerals are ranked along the Mohs scale, which is composed of 10 minerals that have been given arbitrary hardness values. Mohs hardness test, while greatly facilitating the identification of minerals in the field, is not suitable for accurately gauging the hardness of industrial materials such as steel or ceramics. For engineering materials, a variety of instruments have been developed over the years to provide a precise measure of hardness. Many apply a load and measure the depth or size of the resulting indentation. Hardness can be measured on the macro-, micro- or nano- scale.

Brinell Hardness Test

The oldest of the hardness test methods in common use on engineering materials today is the Brinell hardness test. Dr. J. A. Brinell invented the Brinell test in Sweden in 1900. The Brinell test uses a desktop machine to applying a specified load to a hardened sphere of a specified diameter. The Brinell hardness number, or simply the Brinell number, is obtained by dividing the load used, in kilograms, by the measured surface area of the indentation, in square millimeters, left on the test surface. The Brinell test is frequently used to determine the hardness metal forgings and castings that have a large grain structures. The Brinell test provides a measurement over a fairly large area that is less affected by the course grain structure of these materials than are Rockwell or Vickers tests.

A wide range of materials can be tested using a Brinell test simply by varying the test load and indenter ball size. In the USA, Brinell testing is typically done on iron and steel castings using a 3000Kg test force and a 10mm diameter ball. A 1500 kilogram load is usually used for aluminum castings. Copper, brass and thin stock are frequently tested using a 500Kg test force and a 10 or 5mm ball. In Europe Brinell testing is done using a much wider range of forces and ball sizes and it is common to perform Brinell tests on small parts using a 1mm carbide ball and a test force as low as 1kg. These low load tests are commonly referred to as baby Brinell tests. The test conditions should be reported along with the Brinell hardness number. A value reported as "60 HB 10/1500/30" means that a Brinell Hardness of 60 was obtained using a 10mm diameter ball with a 1500 kilogram load applied for 30 seconds.

Rockwell Hardness Test

The Rockwell Hardness test also uses a machine to apply a specific load and then measure the depth of the resulting impression. The indenter may either be a steel ball of some specified diameter or a spherical diamond-tipped cone of 120° angle and 0.2 mm tip radius, called a brale. A minor load of 10 kg is first applied, which causes a small initial penetration to seat the indenter and remove the effects of any surface irregularities. Then, the dial is set to zero and the major load is applied. Upon removal of the major load, the depth reading is taken while the minor load is still on. The hardness number may then be read directly from the scale. The indenter and the test load used determine the hardness scale that is used (A, B, C, etc).

For soft materials such as copper alloys, soft steel, and aluminum alloys a 1/16" diameter steel ball is used with a 100-kilogram load and the hardness is read on the "B" scale. In testing harder materials, hard cast iron and many steel alloys, a 120 degrees diamond cone is used with up to a 150 kilogram load and the hardness is read on the "C" scale. There are several Rockwell scales other than the "B" & "C" scales, (which are called the common scales). A properly reported Rockwell value will have the hardness number followed by "HR" (Hardness Rockwell) and the scale letter. For example, 50 HRB indicates that the material has a hardness reading of 50 on the B scale.

- A -Cemented carbides, thin steel and shallow case hardened steel.
- B -Copper alloys, soft steels, aluminum alloys, malleable iron, etc.
- C -Steel, hard cast irons, pearlitic malleable iron, titanium, deep case hardened steel and other materials harder than B 100.

- D -Thin steel and medium case hardened steel and pearlitic malleable iron.
- E -Cast iron, aluminum and magnesium alloys, bearing metals.
- F -Annealed copper alloys, thin soft sheet metals.
- G -Phosphor bronze, beryllium copper, malleable irons.
- H -Aluminum, zinc, lead K, L, M, P, R, S, V -Bearing metals and other very soft or thin materials, including plastics.

Rockwell Superficial Hardness Test

The Rockwell Superficial Hardness Tester is used to test thin materials, lightly carburized steel surfaces, or parts that might bend or crush under the conditions of the regular test. This tester uses the same indenters as the standard Rockwell tester but the loads are reduced. A minor load of 3 kilograms is used and the major load is either 15 or 45 kilograms depending on the indenter used. Using the 1/16" diameter, steel ball indenter, a "T" is added (meaning thin sheet testing) to the superficial hardness designation. An example of a superficial Rockwell hardness is 23 HR15T, which indicates the superficial hardness as 23, with a load of 15 kilograms using the steel ball.

Vickers and Knoop Microhardness Tests

The Vickers and Knoop Hardness Tests are a modification of the Brinell test and are used to measure the hardness of thin film coatings or the surface hardness of case-hardened parts. With these tests, a small diamond pyramid is pressed into the sample under loads that are much less than those used in the Brinell test. The difference between the Vickers and the Knoop Tests is simply the shape of the diamond pyramid indenter. The Vickers test uses a square pyramidal indenter which is prone to crack brittle materials. Consequently, the Knoop test using a rhombic-based (diagonal ratio 7.114:1) pyramidal indenter was developed which produces longer but shallower indentations. For the same load, Knoop indentations are about 2.8 times longer than Vickers indentations.

An applied load ranging from 10g to 1,000g is used. This low amount of load creates a small indent that must be measured under a microscope. The measurements for hard coatings like TiN must be taken at very high magnification (i.e. 1000X), because the indents are so small. The surface usually needs to be polished. The diagonals of the impression are measured, and these values are used to obtain a hardness number (VHN), usually from a lookup table or chart. The Vickers test can be used to characterize very hard materials but the hardness is measured over a very small region.

The values are expressed like 2500 HK25 (or HV25) meaning 2500 Hardness Knoop at 25 gram force load. The Knoop and Vickers hardness values differ slightly, but for hard coatings, the values are close enough to be within the measurement error and can be used interchangeably.

Scleroscope and Rebound Hardness Tests

The Scleroscope test is a very old test that involves dropping a diamond tipped hammer, which falls inside a glass tube under the force of its own weight from a fixed height, onto the test specimen. The height of the rebound travel of the hammer is measured on a graduated scale. The scale of the rebound is arbitrarily chosen and consists on Shore units, divided into 100 parts, which represent the average rebound from pure hardened high-carbon steel. The scale is continued higher than 100 to include metals having greater hardness. The Shore Scleroscope measures hardness in terms of the elasticity of the material and the hardness number depends on the height to which the hammer rebounds, the harder the material, the higher the rebound.

The Rebound Hardness Test Method is a recent advancement that builds on the Scleroscope. There are a variety of electronic instruments on the market that measure the loss of energy of the impact body. These instruments typically use a spring to accelerate a spherical, tungsten carbide tipped mass towards the surface of the test object. When the mass contacts the surface it has a specific kinetic energy and the impact produces an indentation (plastic deformation) on the surface which takes some of this energy from the impact body. The impact body will lose more energy and it rebound velocity will be less when a larger indentation is produced on softer material. The velocities of the impact body before and after impact are measured and the loss of velocity is related to Brinell, Rockwell, or other common hardness value.

Durometer Hardness Test

A Durometer is an instrument that is commonly used for measuring the indentation hardness of rubbers/elastomers and soft plastics such as polyolefin, fluoropolymer, and vinyl. A Durometer simply uses a calibrated spring to apply a specific pressure to an indenter foot. The indenter foot can be either cone or sphere shaped. An indicating device measures the depth of indentation. Durometers are available in a variety of models and the most popular testers are the Model A used for measuring softer materials and the Model D for harder materials.

Barcol Hardness Test

The Barcol hardness test obtains a hardness value by measuring the penetration of a sharp steel point under a spring load. The specimen is placed under the indenter of the Barcol hardness tester and a uniform pressure is applied until the dial indication reaches a maximum. The Barcol hardness test method is used to determine the hardness of both reinforced and non-reinforced rigid plastics and to determine the degree of cure of resins and plastics.

FRACTURE TOUGHNESS

Fracture toughness is an indication of the amount of stress required to propagate a preexisting flaw. It is a very important material property since the occurrence of flaws is not completely avoidable in the processing, fabrication, or service of a material/component. Flaws may appear as cracks, voids, metallurgical inclusions, weld defects, design discontinuities, or some combination thereof. Since engineers can never be totally sure that a material is flaw free, it is common practice to assume that a flaw of some chosen size will be present in some number of components and use the linear elastic fracture mechanics (LEFM) approach to design critical components. This approach uses the flaw size and features, component geometry, loading conditions and the material property called fracture toughness to evaluate the ability of a component containing a flaw to resist fracture.

A parameter called the stress-intensity factor (K) is used to determine the fracture toughness of most materials. A Roman numeral subscript indicates the mode of fracture and the three modes of fracture are illustrated in the image below. Mode I fracture is the condition in which the crack plane is normal to the direction of largest tensile loading. This is the most commonly encountered mode and, therefore, for the remainder of the material we will consider K_I.

The stress intensity factor is a function of loading, crack size, and structural geometry. The stress intensity factor may be represented by the following equation:

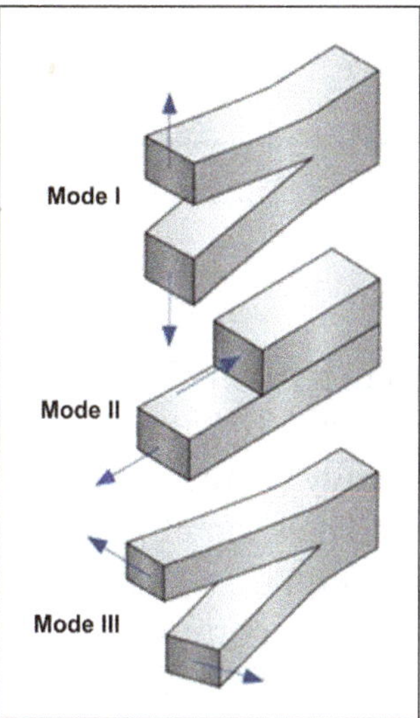

$$K_1 = \sigma\sqrt{\pi a \beta}$$

Where:

KI is the fracture toughness in $MPa\sqrt{m}\,(psi\sqrt{in})$

s is the applied stress in MPa or psi

a is the crack length in meters or inches

B is a crack length and component geometry factor that is different for each specimen and is dimensionless.

Role of Material Thickness

Specimens having standard proportions but different absolute size produce different values for K_I. This results because the stress states adjacent to the flaw changes with the specimen thickness (B) until the thickness exceeds some critical dimension. Once the thickness exceeds the critical dimension, the value of K_I becomes relatively constant and this value, K_{IC}, is a true material property which is called the plane-strain fracture toughness. The relationship between stress intensity, K_I, and fracture toughness, K_{IC}, is similar to the relationship between stress and tensile stress. The stress intensity, K_I, represents the level of "stress" at the tip of the crack and the fracture toughness, K_{IC}, is the highest value of stress intensity that a material under very specific (plane-strain) conditions that a material can withstand without fracture. As the stress intensity factor reaches the K_{IC} value, unstable fracture occurs. As with a material's other mechanical properties, K_{IC} is commonly reported in reference books and other sources.

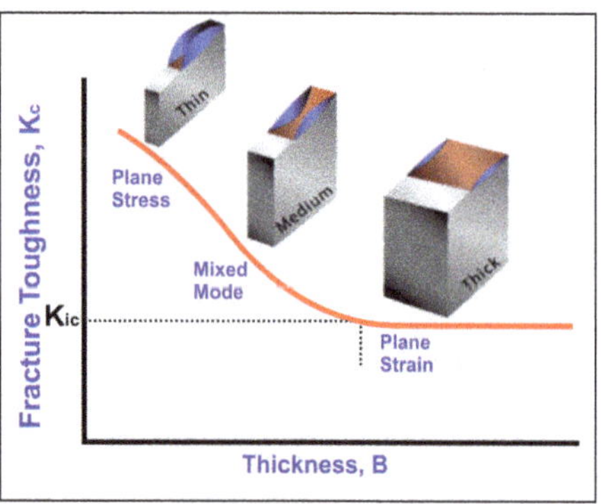

Plane-Strain and Plane-Stress

When a material with a crack is loaded in tension, the materials develop plastic strains as the yield stress is exceeded in the region near the crack tip. Material within the crack tip stress field, situated close to a free surface, can deform laterally (in the z-direction of the image) because there can be no stresses normal to the free surface. The state of stress tends to biaxial and the material fractures in a characteristic ductile manner, with a 45° shear lip being formed at each free surface. This condition is called "plane-stress" and it occurs in relatively thin bodies where the stress through the thickness cannot vary appreciably due to the thin section.

However, material away from the free surfaces of a relatively thick component is not free to deform laterally as it is constrained by the surrounding material. The stress state under these conditions tends to triaxial and there is zero strain perpendicular to both the stress axis and the direction of crack propagation when a material is loaded in tension. This condition is called "plane-strain" and is found in thick plates. Under plane-strain conditions, materials behave essentially elastic until the fracture stress is reached and then rapid fracture occurs. Since little or no plastic deformation is noted, this mode fracture is termed brittle fracture.

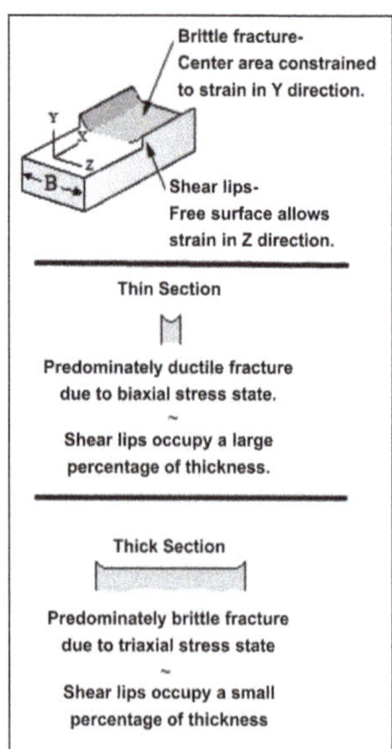

Plane-Strain Fracture Toughness Testing

When performing a fracture toughness test, the most common test specimen configurations are the single edge notch bend (SENB or three-point bend), and the compact tension (CT) specimens. It is clear that an accurate determination of the plane-strain fracture toughness requires a specimen whose thickness exceeds some critical thickness (B). Testing has shown that plane-strain conditions generally prevail when:

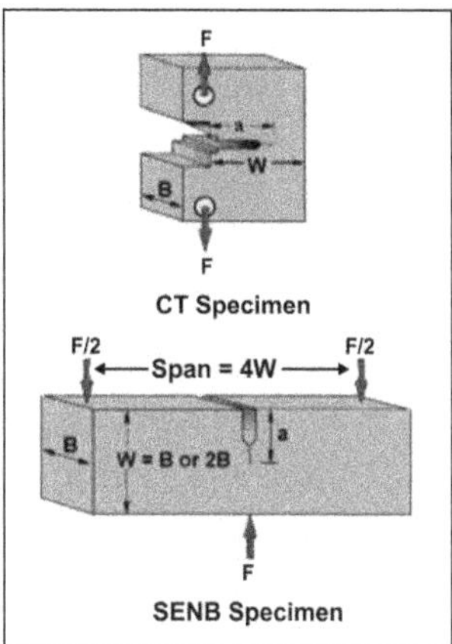

$$B \geq 2.5\left(\frac{K_{IC}}{\sigma y}\right)^2$$

Where, B is the minimum thickness that produces a condition where plastic strain energy at the crack tip in minimal

- K_{IC} is the fracture toughness of the material.
- S_y is the yield stress of material.

When a material of unknown fracture toughness is tested, a specimen of full material section thickness is tested or the specimen is sized based on a prediction of the fracture toughness. If the fracture toughness value resulting from the test does not satisfy the requirement of the above equation, the test must be repeated using a thicker specimen. In addition to this thickness calculation, test specifications have several other requirements that must be met (such as the size of the shear lips) before a test can be said to have resulted in a K_{IC} value.

When a test fails to meet the thickness and other test requirement that are in place to insure plane-strain condition, the fracture toughness values produced is given the designation K_C. Sometimes it is not possible to produce a specimen that meets the thickness requirement. For example when a relatively thin plate product with high toughness is being tested, it might not be possible to produce a thicker specimen with plain-strain conditions at the crack tip.

Plane-Stress and Transitional-Stress States

For cases where the plastic energy at the crack tip is not negligible, other fracture mechanics parameters, such as the J integral or R-curve, can be used to characterize a material. The toughness data produced by these other tests will be dependant on the thickness of the product tested and will not be a true material property. However, plane-strain conditions do not exist in all structural configurations and using K_{IC} values in the design of relatively thin areas may result in excess conservatism and a weight or cost penalty. In cases where the actual stress state is plane-stress or, more generally, some intermediate- or transitional-stress state, it is more appropriate to use J integral or R-curve data, which account for slow, stable fracture (ductile tearing) rather than rapid (brittle) fracture.

Uses of Plane-Strain Fracture Toughness

K_{IC} values are used to determine the critical crack length when a given stress is applied to a component.

$$\sigma_C \leq \frac{K_{IC}}{Y\sqrt{\pi a}}$$

Where

s_c is the critical applied stress that will cause failure

K_{IC} is the plane-strain fracture toughness

Y is a constant related to the sample's geometry

a is the crack length for edge cracks or one half crack length for internal crack

K_{IC} values are used also used to calculate the critical stress value when a crack of a given length is found in a component.

$$a_c = \frac{1}{\pi}\left(\frac{K_{IC}}{\sigma Y}\right)^2$$

Where:

a is the crack length for edge cracks or one half crack length for internal crack

s is the stress applied to the material

K_{IC} is the plane-strain fracture toughness

Y is a constant related to the sample's geometry

Orientation

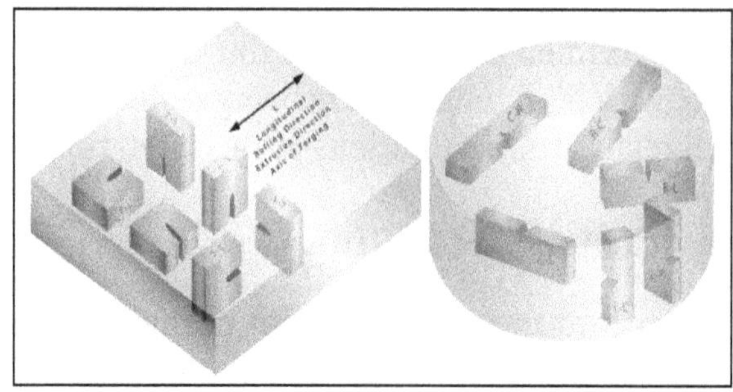

The fracture toughness of a material commonly varies with grain direction. Therefore, it is customary to specify specimen and crack orientations by an ordered pair of grain direction symbols. The first letter designates the grain direction normal to the crack plane. The second letter designates the grain direction parallel to the fracture plane. For flat sections of various products, e.g., plate, extrusions, forgings, etc., in which the three grain directions are designated (L) longitudinal, (T) transverse, and (S) short transverse, the six principal fracture path directions are: L-T, L-S, T-L, T-S, S-L and S-T.

CREEP AND STRESS RUPTURE PROPERTIES

Creep is a time-dependent deformation of a material while under an applied load that is below its yield strength. It is most often occurs at elevated temperature, but some materials creep at room temperature. Creep terminates in rupture if steps are not taken to bring to a halt.

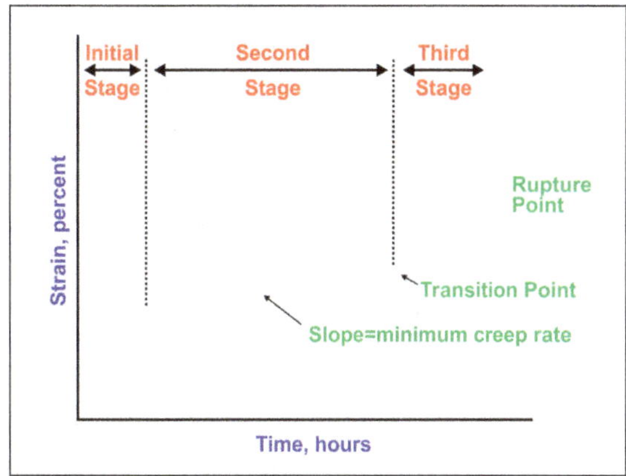

Creep data for general design use are usually obtained under conditions of constant uniaxial loading and constant temperature. Results of tests are usually plotted as strain versus time up to rupture. As indicated in the image, creep often takes place in three stages. In the initial stage, strain occurs at a relatively rapid rate but the rate gradually decreases until it becomes approximately constant during the second stage. This constant creep rate is called the minimum creep rate or steady-state creep rate since it is the slowest creep rate during the test. In the third stage, the strain rate increases until failure occurs.

Creep in service is usually affected by changing conditions of loading and temperature and the number of possible stress-temperature-time combinations is infinite. While most materials are subject to creep, the creep mechanisms is often different between metals, plastics, rubber, concrete.

Stress Rupture Properties

Stress rupture testing is similar to creep testing except that the stresses are higher than those used in a creep testing. Stress rupture tests are used to determine the time necessary to produce failure so stress rupture testing is always done until failure. Data is plotted log-log. A straight line or best fit curve is usually obtained at each temperature of interest. This information can then be used to extrapolate time to failure for longer times. A typical set of stress rupture curves is shown below.

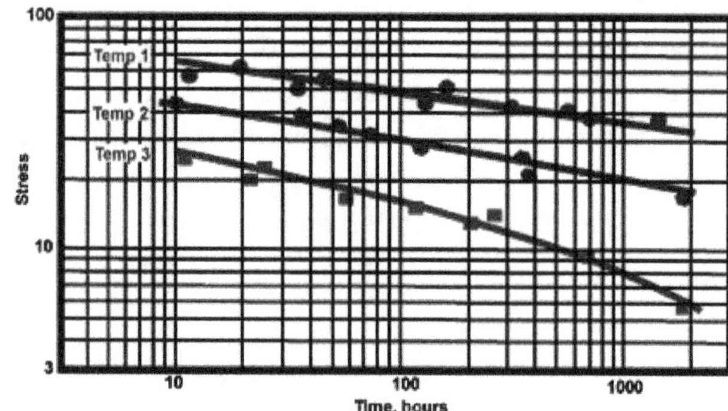

STRENGTHENING MECHANISMS OF MATERIALS

Methods have been devised to modify the yield strength, ductility, and toughness of both crystalline and amorphous materials. These strengthening mechanisms give engineers the ability to tailor the mechanical properties of materials to suit a variety of different applications. For example, the favorable properties of steel result from interstitial incorporation of carbon into the iron lattice. Brass, a binary alloy of copper and zinc, has superior mechanical properties compared to its constituent metals due to solution strengthening. Work hardening (such as beating a red-hot piece of metal on anvil) has also been used for centuries by blacksmiths to introduce dislocations into materials, increasing their yield strengths.

Plastic deformation occurs when large numbers of dislocations move and multiply so as to result in macroscopic deformation. In other words, it is the movement of dislocations in the material which allows for deformation. If we want to enhance a material's mechanical properties (i.e. increase the yield and tensile strength), we simply need to introduce a mechanism which prohibits the mobility of these dislocations. Whatever the mechanism may be, (work hardening, grain size reduction, etc.) they all hinder dislocation motion and render the material stronger than previously.

The stress required to cause dislocation motion is orders of magnitude lower than the theoretical stress required to shift an entire plane of atoms, so this mode of stress relief is energetically favorable. Hence, the hardness and strength (both yield and tensile) critically depend on the ease with which dislocations move. Pinning points, or locations in the crystal that oppose the motion of dislocations, can be introduced into the lattice to reduce dislocation mobility, thereby increasing mechanical strength. Dislocations may be pinned due to stress field interactions with other dislocations and solute particles, creating physical barriers from second phase precipitates forming along grain boundaries. There are four main strengthening mechanisms for metals, each is a method to prevent dislocation motion and propagation, or make it energetically unfavorable for the dislocation to move. For a material that has been strengthened, by some processing method, the amount of force required to start irreversible (plastic) deformation is greater than it was for the original material.

In amorphous materials such as polymers, amorphous ceramics (glass), and amorphous metals, the lack of long range order leads to yielding via mechanisms such as brittle fracture, crazing, and shear band formation. In these systems, strengthening mechanisms do not involve dislocations, but rather consist of modifications to the chemical structure and processing of the constituent material.

The strength of materials cannot infinitely increase. Each of the mechanisms explained below involves some trade-off by which other material properties are compromised in the process of strengthening.

Strengthening Mechanisms in Metals

Work Hardening

The primary species responsible for work hardening are dislocations. Dislocations interact with each other by generating stress fields in the material. The interaction between the stress fields of dislocations can impede dislocation motion by repulsive or attractive interactions. Additionally, if two dislocations cross, dislocation line entanglement occurs, causing the formation of a jog which opposes dislocation motion. These entanglements and jogs act as pinning points, which oppose dislocation motion. As both of these processes are more likely to occur when more dislocations are present, there is a correlation between dislocation density and yield strength,

$$\Delta\sigma_y = Gb\sqrt{\rho_\perp}$$

where G is the shear modulus, b is the Burgers vector, and $\rho_\perp$ is the dislocation density.

Increasing the dislocation density increases the yield strength which results in a higher shear stress required to move the dislocations. This process is easily observed while working a material (in metals cold working of process). Theoretically, the strength of a material with no dislocations will be extremely high (τ=G/2) because plastic deformation would require the breaking of many bonds simultaneously. However, at moderate dislocation density values of around 10^7-10^9 dislocations/ m^2, the material will exhibit a significantly lower mechanical strength. Analogously, it is easier to move a rubber rug across a surface by propagating a small ripple through it than by dragging the whole rug. At dislocation densities of 10^{14} dislocations/m^2 or higher, the strength of the material becomes high once again. Also, the dislocation density cannot be infinitely high, because then the material would lose its crystalline structure.

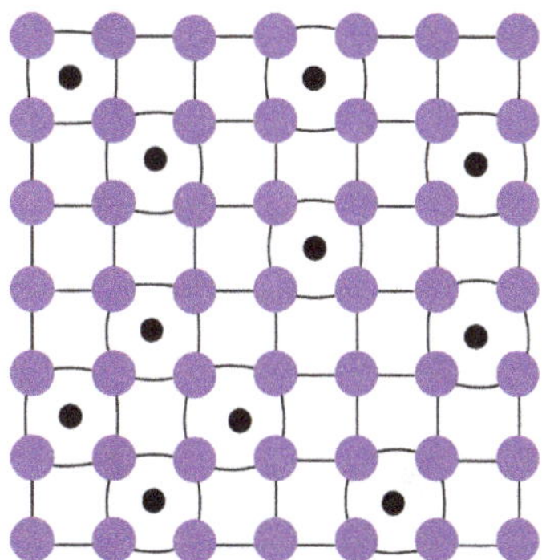

In figure, this is a schematic illustrating how the lattice is strained by the addition of interstitial solute. Notice the strain in the lattice that the solute atoms cause. The interstitial solute could be carbon in iron for example. The carbon atoms in the interstitial sites of the lattice creates a stress field that impedes dislocation movement.

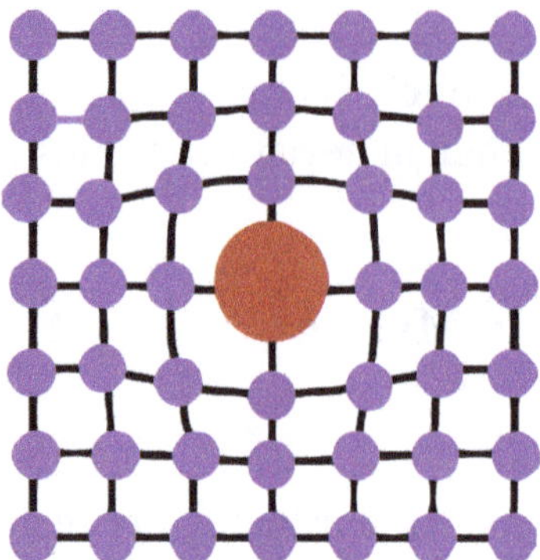

This is a schematic illustrating how the lattice is strained by the addition of substitutional solute. Notice the strain in the lattice that the solute atom causes.

Solid Solution Strengthening and Alloying

For this strengthening mechanism, solute atoms of one element are added to another, resulting in either substitutional or interstitial point defects in the crystal. The solute atoms cause lattice distortions that impede dislocation motion, increasing the yield stress of the material. Solute atoms have stress fields around them which can interact with those of dislocations. The presence of solute atoms impart compressive or tensile stresses to the lattice, depending on solute size, which interfere with nearby dislocations, causing the solute atoms to act as potential barriers.

The shear stress required to move dislocations in a material is:

$$\Delta\tau = Gb\sqrt{c}\epsilon^{3/2}$$

where c is the solute concentration and ϵ is the strain on the material caused by the solute.

Increasing the concentration of the solute atoms will increase the yield strength of a material, but there is a limit to the amount of solute that can be added, and one should look at the phase diagram for the material and the alloy to make sure that a second phase is not created.

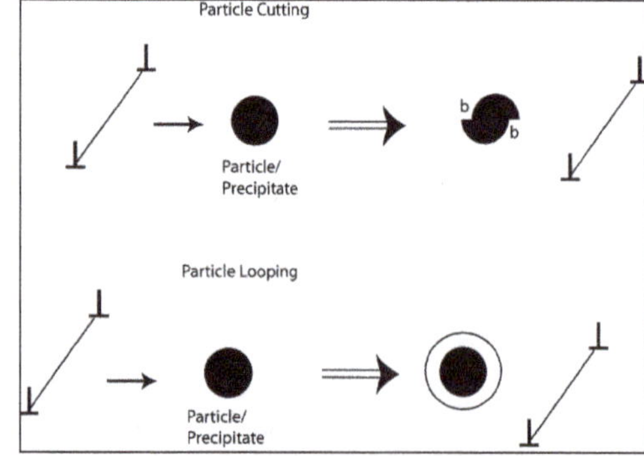

A schematic illustrating how the dislocations can interact with a particle. It can either cut through the particle or bow around the particle and create a dislocation loop as it moves over the particle.

In general, the solid solution strengthening depends on the concentration of the solute atoms, shear modulus of the solute atoms, size of solute atoms, valency of solute atoms (for ionic materials), and the symmetry of the solute stress field. The magnitude of strengthening is higher for non-symmetric stress fields because these solutes can interact with both edge and screw dislocations, whereas symmetric stress fields, which cause only volume change and not shape change, can only interact with edge dislocations.

Precipitation Hardening

In most binary systems, alloying above a concentration given by the phase diagram will cause the formation of a second phase. A second phase can also be created by mechanical or thermal treatments. The particles that compose the second phase precipitates act as pinning points in a similar manner to solutes, though the particles are not necessarily single atoms.

The dislocations in a material can interact with the precipitate atoms in one of two ways. If the precipitate atoms are small, the dislocations would cut through them. As a result, new surfaces of the particle would get exposed to the matrix and the particle-matrix interfacial energy would increase. For larger precipitate particles, looping or bowing of the dislocations would occur and result in dislocations getting longer. Hence, at a critical radius of about 5 nm, dislocations will preferably cut across the obstacle, while for a radius of 30 nm, the dislocations will readily bow or loop to overcome the obstacle.

The mathematical descriptions are as follows:

For particle bowing- $\Delta\tau = \frac{Gb}{L-2r}$

For particle cutting- $\Delta\tau = \frac{\gamma\pi r}{bL}$

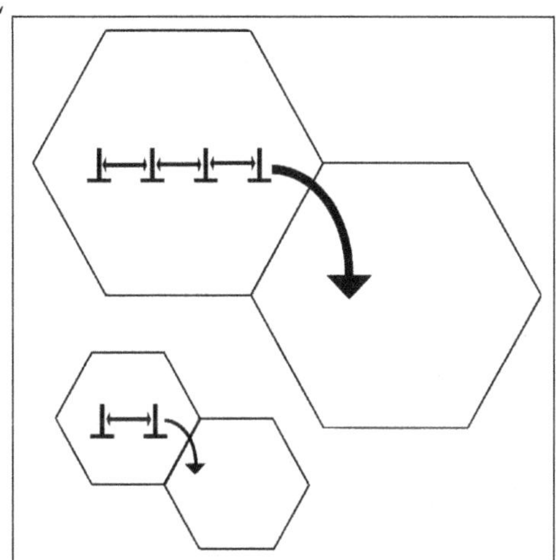

A schematic roughly illustrating the concept of dislocation pile up and how it effects the strength of the material. A material with larger grain size is able to have more dislocation to pile up leading to a bigger driving force for dislocations to move from one grain to another. Thus, less force need be applied to move a dislocation from a larger, than from a smaller grain, leading materials with smaller grains to exhibit higher yield stress.

Grain Boundary Strengthening

In a polycrystalline metal, grain size has a tremendous influence on the mechanical properties. Because grains usually have varying crystallographic orientations, grain boundaries arise. While undergoing deformation, slip motion will take place. Grain boundaries act as an impediment to dislocation motion for the following two reasons:

1. Dislocation must change its direction of motion due to the differing orientation of grains.
2. Discontinuity of slip planes from grain one to grain two.

The stress required to move a dislocation from one grain to another in order to plastically deform a material depends on the grain size. The average number of dislocations per grain decreases with average grain size. A lower number of dislocations per grain results in a lower dislocation 'pressure' building up at grain boundaries. This makes it more difficult for dislocations to move into adjacent grains. This relationship is the Hall-Petch relationship and can be mathematically described as follows:

$$\sigma_y = \sigma_{y,0} + \frac{k}{d^x},$$

where k is a constant, d is the average grain diameter and $\sigma_{y,0}$ is the original yield stress.

The fact that the yield strength increases with decreasing grain size is accompanied by the caveat that the grain size cannot be decreased infinitely. As the grain size decreases, more free volume is generated resulting in lattice mismatch. Below approximately 10 nm, the grain boundaries will tend to slide instead; a phenomenon known as grain-boundary sliding. If the grain size gets too small, it becomes more difficult to fit the dislocations in the grain and the stress required to move them is less. It was not possible to produce materials with grain sizes below 10 nm until recently, so the discovery that strength decreases below a critical grain size is still finding new applications.

Transformation Hardening

This method of hardening is used for steels.

High-strength steels generally fall into three basic categories, classified by the strengthening mechanism employed. 1- solid-solution-strengthened steels (rephos steels) 2- grain-refined steels or high strength low alloy steels (HSLA) 3- transformation-hardened steels

Transformation-hardened steels are the third type of high-strength steels.These steels use predominantly higher levels of C and Mn along with heat treatment to increase strength. The finished product will have a duplex micro-structure of ferrite with varying levels of degenerate martensite. This allows for varying levels of strength. There are three basic types of transformation-hardened steels. These are dual-phase (DP), transformation-induced plasticity (TRIP), and martensitic steels.

The annealing process for dual -phase steels consists of first holding the steel in the alpha + gamma temperature region for a set period of time. During that time C and Mn diffuse into the austenite leaving a ferrite of greater purity. The steel is then quenched so that the austenite is transformed into martensite, and the ferrite remains on cooling. The steel is then subjected to a temper cycle to allow some level of marten-site decomposition. By controlling the amount of martensite in the steel, as well as the degree of temper, the strength level can be controlled. Depending on processing and chemistry, the strength level can range from 350 to 960 MPa.

TRIP steels also use C and Mn, along with heat treatment, in order to retain small amounts of Austen and bainite in a ferrite matrix. Thermal processing for TRIP steels again involves annealing the steel in the a + g region for a period of time sufficient to allow C and Mn to diffuse into austenite. The steel is then quenched to a point above the martensite start temperature and held there. This allows the formation of bainite, an austenite decomposition product. While at this temperature, more C is allowed to enrich the retained austenite. This, in turn, lowers the martensite start temperature to below room temperature. Upon final quenching a metastable austenite is retained in the predominantly ferrite matrix along with small amounts of bainite (and other forms of decomposed austenite). This combination of micro-structures has the added benefits of higher strengths and resistance to necking during forming. This offers great improvements in formability over other high-strength steels. Essentially, as the TRIP steel is being formed, it becomes much stronger. Tensile strengths of TRIP steels are in the range of 600-960 MPa.

Martensitic steels are also high in C and Mn. These are fully quenched to martensite during processing. The martensite structure is then tempered back to the appropriate strength level, adding toughness to the steel. Tensile strengths for these steels range as high as 1500 MPa.

Strengthening Mechanisms in Amorphous Materials

Polymer

Polymers fracture via breaking of inter- and intra molecular bonds; hence, the chemical structure of these materials plays a huge role in increasing strength. For polymers consisting of chains which easily slide past each other, chemical and physical cross linking can be used to increase rigidity and yield strength. In thermoset polymers (thermosetting plastic), disulfide bridges and other covalent cross links give rise to a hard structure which can withstand very high temperatures. These cross-links are particularly helpful in improving tensile strength of materials which contain lots of free volume prone to crazing, typically glassy brittle polymers. In thermoplastic elastomer, phase separation of dissimilar monomer components leads to association of hard domains within a sea of soft phase, yielding a physical structure with increased strength and rigidity. If yielding occurs by chains sliding past each other (shear bands), the strength can also be increased by introducing kinks into the polymer chains via unsaturated carbon-carbon bonds.

Adding filler materials such as fibers, platelets, and particles is a commonly employed technique for strengthening polymer materials. Fillers such as clay, silica, and carbon network materials have been extensively researched and used in polymer composites in part due to their effect on mechanical properties. Stiffness-confinement effects near rigid interfaces, such as those between a polymer matrix and stiffer filler materials, enhance the stiffness of composites by restricting polymer chain motion. This is especially present where fillers are chemically treated to strongly interact with polymer chains, increasing the anchoring of polymer chains to the filler interfaces and thus further restricting the motion of chains away from the interface. Stiffness-confinement effects have been characterized in model nanocomposites, and shows that composites with length scales on the order of nanometers increase the effect of the fillers on polymer stiffness dramatically.

Increasing the bulkiness of the monomer unit via incorporation of aryl rings is another strengthening mechanism. The anisotropy of the molecular structure means that these mechanisms are heavily dependent on the direction of applied stress. While aryl rings drastically increase rigidity along the direction of the chain, these materials may still be brittle in perpendicular directions. Macroscopic structure can be adjusted to compensate for this anisotropy. For example, the high strength of Kevlar arises from a stacked multilayer macrostructure where aromatic polymer layers are rotated with respect to their neighbors. When loaded oblique to the chain direction, ductile polymers with flexible linkages, such as oriented polyethylene, are highly prone to shear band formation, so macroscopic structures which place the load parallel to the draw direction would increase strength.

Mixing polymers is another method of increasing strength, particularly with materials that show crazing preceding brittle fracture such as atactic polystyrene (APS). For example, by forming a 50/50 mixture of APS with polyphenylene oxide (PPO), this embrittling tendency can be almost completely suppressed, substantially increasing the fracture strength.

Interpenetrating polymer networks (IPNs), consisting of interlacing crosslinked polymer networks that are not covalently bonded to one another, can lead to enhanced strength in polymer materials. The use of an IPN approach imposes compatibility (and thus macroscale homogeneity) on otherwise immiscible blends, allowing for a blending of mechanical properties. For example, silicone-polyurethane IPNs show increased tear and flexural strength over base silicone networks, while preserving the high elastic recovery of the silicone network at high strains. Increased stiffness can also be achieved by pre-straining polymer networks and then sequentially forming a secondary network within the strained material. This takes advantage of the anisotropic strain hardening of the original network (chain alignment from stretching of the polymer chains) and provides a mechanism whereby the two networks transfer stress to one another due to the imposed strain on the pre-strained network.

Glass

Many silicate glasses are strong in compression but weak in tension. By introducing compression stress into the structure, the tensile strength of the material can be increased. This is typically done via two mechanisms: thermal treatment (tempering) or chemical bath (via ion exchange).

In tempered glasses, air jets are used to rapidly cool the top and bottom surfaces of a softened (hot) slab of glass. Since the surface cools quicker, there is more free volume at the surface than in the bulk melt. The core of the slab then pulls the surface inward, resulting in an internal compressive stress at the surface. This substantially increases the tensile strength of the material as tensile stresses exerted on the glass must now resolve the compressive stresses before yielding.

$$\sigma_{y=modified} = \sigma_{y,0} + \sigma_{compressive}$$

Alternately, in chemical treatment, a glass slab treated containing network formers and modifiers is submerged into a molten salt bath containing ions larger than those present in the modifier. Due to a concentration gradient of the ions, mass transport must take place. As the larger cation diffuses from the molten salt into the surface, it replaces the smaller ion from the modifier. The larger ion squeezing into surface introduces compressive stress in the glass's surface. A common example is treatment of sodium oxide modified silicate glass in molten potassium chloride.

Composite Strengthening

Many of the basic strengthening mechanisms can be classified based on their dimensionality. At 0-D there is precipitate and solid solution strengthening with particulates strengthening structure, at 1-D there is work/forest hardening with line dislocations as the hardening mechanism, and at 2-D there is grain boundary strengthening with surface energy of granular interfaces providing strength improvement. The two primary types of composite strengthening, fiber reinforcement and laminar reinforcement, fall in the 1-D and 2-D classes, respectively. The anisotropy of fiber and laminar composite strength reflects these dimensionalities. The primary idea behind composite strengthening is to combine materials with opposite strengths and weaknesses to create a material which transfers load onto the stiffer material but benefits from the ductility and toughness of the softer material.

Fiber Reinforcement

Fiber-reinforced composites (FRCs) consist of a matrix of one material containing parallel embedded fibers. There are two variants of fiber-reinforced composites, one with stiff fibers and a ductile matrix and one with ductile fibers and a stiff matrix. The former variant is exemplified by fiberglass which contains very strong but delicate glass fibers embedded in a softer plastic matrix resilient to fracture. The latter variant is found in almost all buildings as reinforced concrete with ductile, high tensile-strength steel rods embedded in brittle, high compressive-strength concrete. In both cases, the matrix and fibers have complimentary mechanical properties and the resulting composite material is therefore more practical for applications in the real world.

For a composite containing aligned, stiff fibers which span the length of the material and a soft, ductile matrix, the following descriptions provide a rough model.

Four Stages of Deformation

The condition of a fiber-reinforced composite under applied tensile stress along the direction of the fibers can be decomposed into four stages from small strain to large strain. Since the stress is parallel to the fibers, the deformation is described by the isostrain condition, i.e., the fiber and matrix experience the same strain. At each stage, the composite stress (σ_c) is given in terms of the volume fractions of the fiber and matrix (V_f, V_m), the Young's moduli of the fiber and matrix (E_f, E_m), the strain of the composite (ϵ_c), and the stress of the fiber and matrix as read from a stress-strain curve ($\sigma_f(\epsilon_c), \sigma_m(\epsilon_c)$).

1. Both fiber and composite remain in the elastic strain regime. In this stage, we also note that the composite Young's modulus is a simple weighted sum of the two component moduli:

$$\sigma_c = V_f \epsilon_c E_f + V_m \epsilon_c E_m$$

$$E_c = V_f E_f + V_m E_m$$

2. The fiber remains in the elastic regime but the matrix yields and plastically deforms:

$$\sigma_c = V_f \epsilon_c E_f + V_m \sigma_m (\epsilon_c)$$

3. Both fiber and composite yield and plastically deform. This stage often features significant Poisson strain which is not captured by model below:

$$\sigma_c = V_f \sigma_f (\epsilon_c) + V_m \sigma_m (\epsilon_c)$$

4. The fiber fractures while the matrix continues to plastically deform. While in reality the fractured pieces of fiber still contribute some strength, it is left out of this simple model:

$$\sigma_c \approx V_m \sigma_m (\epsilon_c)$$

Tensile Strength

Due to the heterogeneous nature of FRCs, they also feature multiple tensile strengths (TS), one corresponding to each component. Given the assumptions outlined above, the first tensile strength would correspond to failure of the fibers, with some support from the matrix plastic deformation strength, and the second with failure of the matrix:

$$TS_1 = V_f TS_f + V_m \sigma_m (\epsilon_c)$$

$$TS_2 = V_m TS_m$$

Anisotropy (Orientation Effects)

As a result of the aforementioned dimensionality (1-D) of fiber reinforcement, significant anisotropy is observed in its mechanical properties. The following equations model the tensile strength of a FRC as a function of the misalignment angle (θ) between the fibers and the applied force, the stresses in the parallel and perpendicular, or $\theta = 0$ and 90° o, cases ($\sigma_{\parallel}, \sigma_{\perp}$), and the shear strength of the matrix (τ_{my}).

Small Misalignment Angle (Longitudinal Fracture)

The angle is small enough to maintain load transfer onto the fibers and prevent delamination of fibers *and* the misaligned stress samples a slightly larger cross-sectional area of the fiber so the strength of the fiber is not just maintained but actually increases compared to the parallel case.

$$TS(\theta) = \frac{\sigma_{\parallel}}{\cos^2(\theta)}$$

Significant Misalignment Angle (Shear Failure)

The angle is large enough that the load is not effectively transferred to the fibers and the matrix experiences enough strain to fracture.

$$TS(\theta) = \frac{\tau_{my}}{\sin(\theta)\cos(\theta)}$$

Near Perpendicular Misaligment Angle (Transverse Fracture)

The angle is close to 900 so most of the load remains in the matrix and thus tensile transverse matrix fracture is the dominant failure condition. This can be seen as complementary to the small angle case, with similar form but with an angle $90-\theta$.

$$TS(\theta) = \frac{\sigma_{\perp}}{\sin^2(\theta)}$$

Applications and Current Research

Strengthening of materials is useful in many applications. A primary application of strengthened materials is for construction. In order to have stronger buildings and bridges, one must have a strong frame that can support high tensile or compressive load and resist plastic deformation. The steel frame used to make the building should be as strong as possible so that it does not bend under the entire weight of the building. Polymeric roofing materials would also need to be strong so that the roof does not cave in when there is build-up of snow on the rooftop.

Research is also currently being done to increase the strength of metallic materials through the addition of polymer materials such as bonded carbon fiber reinforced polymer to (CFRP).

Molecular Dynamics Simulations

The use of computation simulations to model work hardening in materials allows for the direct observation of critical elements that rule the process of strengthening materials. The basic reasoning derives from the fact that, when examining plasticity and the movement of dislocations in materials, a focus on the atomistic level is many times not accounted for and the focus rests on the contiuum description of materials. Since the practice of tracking these atomistic effects in experiments and theorizing about them in textbooks cannot provide a full understanding of these interactions, many turn to molecular dynamics simulations to develop this understanding.

The simulations work by utilizing the known atomic interactions between any two atoms and the relationship F = ma, so that the dislocations moving through the material are ruled by simple mechanical actions and reactions of the atoms. The interatomic potential usually utilized to estimate these interactions is the Lennard – Jones 12:6 potential. Lennard – Jones is widely accepted because its experimental shortcomings are well-known. These interactions are simply scaled up to millions or billions of atoms in some cases to simulate materials more accurately.

Molecular dynamic simulations display the interactions based upon the governing equations provided above for the strengthening mechanisms. They provide an effective way to see these mechanisms in action outside the painstaking realm of direct observation during experiments.

FAILURE THEORIES

Theories of failure are those theories which help us to determine the safe dimensions of a machine component when it is subjected to combined stresses due to various loads acting on it during its functionality.

Some examples of such components are as follows:

1. I.C. engine crankshaft.
2. Shaft used in power transmission.
3. Spindle of a screw jaw.
4. Bolted and welded joints used under eccentric loading.
5. Ceiling fan rod.

Theories of failure are employed in the design of a machine component due to the unavailability of failure stresses under combined loading conditions.

Theories of failure play a key role in establishing the relationship between stresses induced under combined loading conditions and properties obtained from tension test like ultimate tensile strength (S_{ut}) and yield strength (S_{yt}).

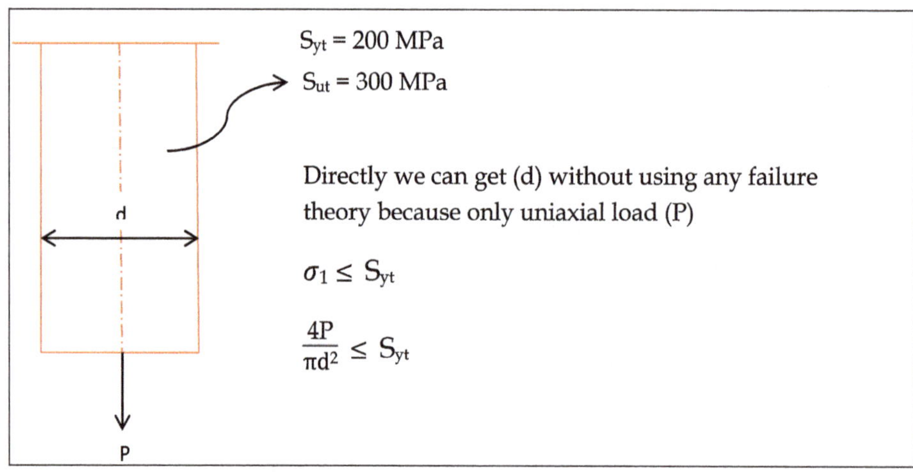

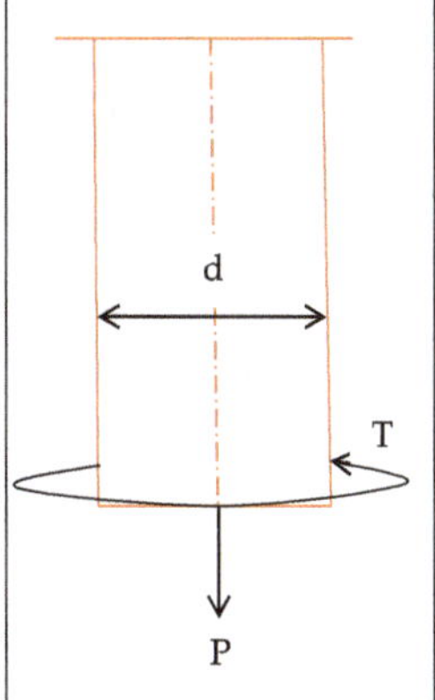

Member is subjected to both Twisting moment and uniaxial load, hence combined loading conditions.

We cannot determine (d) directly in this case because failure stresses under combined loading conditions are unknown.

So, different scientists give relationships between Stresses induced under combined loading conditions and (S_{yt} and S_{ut}) obtained using tension test which are called theories of failure.

Various Theories of Failure

1. Maximum Principal Stress theory also known as Rankine's Theory.
2. Maximum Shear Stress theory or Guest and Tresca's Theory.
3. Maximum Principal Strain theory also known as St. Venant's Theory.
4. Total Strain Energy theory or Haigh's Theory.
5. Maximum Distortion Energy theory or Vonmises and Hencky's Theory.

Maximum Principal Stress Theory (M.P.S.T)

According to M.P.S. T Condition for failure is, Maximum principal stress $(\sigma_1) >$ failure stresses (Syt or Sut) and Factor of safety (F.O.S) = 1 If (σ_1) is +ve then Syt or Sut ( ) is –ve then S_{yc} or S_{uc}.

Condition for Safe Design

Factor of safety (F.O.S) > 1

Maximum principal stress $(\sigma_1) \leq$ Permissible stress (σ per)

$$\text{where permissible stress} = \frac{\text{Failure stress}}{\text{Factor of safety}} = \frac{S_{yt}}{N} \text{or} \frac{S_{ut}}{N}$$

$$\sigma_1 \leq \frac{S_{yt}}{N} or \frac{S_{ut}}{N} N$$

Maximum Shear Stress Theory (M.S.S.T)

Condition for failure, Maximum shear stress induced at a critical point under triaxial combined stress >Yield strength in shear under tensile test.

$$\text{Absolute } \tau_{max} > \underbrace{(S_{ys})T.T}_{\text{unknown therefore use } S_{yt}} \ \ or \ \ \frac{S_{yt}}{2}$$

Condition for Safe Design

Maximum shear stress induced at a critical tensile point under tri-axial combined stress ≤ Permissible shear stress (τ per).

where,

$$\text{Permissible shear stress} = \frac{\text{Yield strength in shear under tension test}}{\text{Factor of safety}} = \frac{(S_{ys})T.T}{N} = \frac{S_{yt}}{2N}$$

$$\text{Absolute } \tau \text{ max} \leq \frac{(S_{ys})T.T}{N} \text{ or } \frac{S_{yt}}{2N}$$

For tri-axial state of stress,

$$\text{larger of } \left[\left|\frac{\sigma_1 - \sigma_2}{2}\right|, \left|\frac{\sigma_2 - \sigma_3}{2}\right|, \left|\frac{\sigma_3 - \sigma_1}{2}\right|\right] \leq \frac{S_{yt}}{2N}$$

$$\text{larger of} \left[\, |\sigma_1 - \sigma_2|, \, |\sigma_2 - \sigma_3|, \, |\sigma_3 - \sigma_1|\right] \leq \frac{S_{yt}}{N}$$

For Biaxial state of stress, $\sigma_3 = 0$

$$\left|\frac{\sigma_1}{2}\right| \ or \ \left|\frac{\sigma_1 - \sigma_2}{2}\right| \leq \frac{S_{yt}}{2N}$$

$$|\sigma_1| \leq \frac{S_{yt}}{N} \text{ when } \sigma_1, \sigma_2 \text{ are like in nature}$$

$$|\sigma_1 - \sigma_2| \leq \frac{S_{yt}}{N} \text{ when } \sigma_1, \sigma_2 \text{ are unlike in nature}$$

Maximum Principal Strain Theory (M.P.St.T)

Condition for failure, Maximum Principal strain (ε_1) > Yielding strain under tensile test $(\varepsilon_{Y.P.})_{T.T}$:

$$\varepsilon 1 > (\varepsilon_{Y.P.})T.T \ or \ \frac{S_{yt}}{N}$$

where E is Young's Modulus of Elasticity

Condition for Safe Design

Maximum Principal strain ≤ Permissible strain

$$\text{where Permissible strain} = \frac{\text{Yielding strain under tensile test}}{\text{Factor of safety}} = \frac{(\varepsilon_{Y.P.})_{T.T}}{N} = \frac{S_{yt}}{EN}$$

$$\varepsilon_1 \le \frac{S_{yt}}{EN}$$

$$\frac{1}{E}\left[\sigma_1 - \mu(\sigma_2 + \sigma_3)\right] \le \frac{S_{yt}}{EN}$$

$$\sigma_1 - \mu(\sigma_2 + \sigma_3) \le \frac{S_{yt}}{N}$$

$$\sigma_1 - \mu(\sigma_2) \le \frac{S_{yt}}{N}$$

Total Strain Energy Theory (T.St.E.T)

Condition for failure, Total Strain Energy per unit volume > Strain energy per unit volume at yield point under tension test.

Condition for Safe Design

Total Strain Energy per unit volume ≤ Strain energy per unit volume at yield point under tension test.

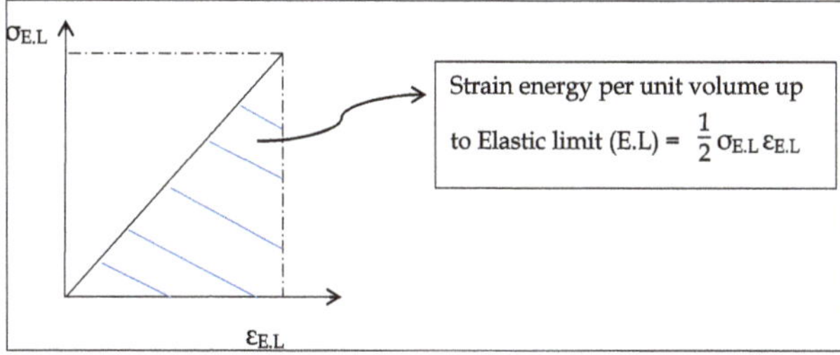

$$\text{Total Strain Energy per unit volume(triaxial)} = \frac{1}{2}\sigma_1\varepsilon_1 + \frac{1}{2}\sigma_2\varepsilon_2 + \frac{1}{2}\sigma_3\varepsilon_3$$

$$\varepsilon_1 = \frac{1}{E}\left[\sigma_1 - \mu(\sigma_2 + \sigma_3)\right]$$

$$\varepsilon_2 = \frac{1}{E}\left[\sigma_2 - \mu(\sigma_1 + \sigma_3)\right]$$

$$\varepsilon_3 = \frac{1}{E}\left[\sigma_3 - \mu(\sigma_1 + \sigma_2)\right]$$

By substituting equations $\frac{1}{2}\ \sigma_1\varepsilon_1 + \frac{1}{2}\sigma_2\ \varepsilon_2 + \frac{1}{2}\ \sigma_3\varepsilon_3$ in equations

Total Strain Energy per unit volume $\leq$ Strain energy per unit volume at yield point under tension test.

$$T.S.E.\ /vol\ =\ \frac{1}{2E}\left[\sigma_1^2 + \sigma_2^{\ 2}\ +\ \sigma_3^2\ -\ 2\mu\left(\sigma_1\ \sigma_2 + \sigma_2\ \sigma_3 + \sigma_3\ \sigma_1\right)\right]$$

To get $[(S.E\ /vol)_{Y.P.}]_{T.T}$,

Substitute $\sigma_1 = \sigma = \frac{S_{yt}}{N}$, $\sigma_2 = \ \sigma_3 = 0$ in equation

$$[(S.E\ /\ vol)_{Y.P.}]_{T.T} = \frac{1}{2E}(\frac{S_{yt}}{N})^{\wedge 2}$$

By Substituting equations $T.S.E.\ /vol\ =\ \frac{1}{2E}\left[\sigma_1^2 + \sigma_2^{\ 2}\ +\ \sigma_3^2\ -\ 2\mu\left(\sigma_1\ \sigma_2 + \sigma_2\ \sigma_3 + \sigma_3\ \sigma_1\right)\right]$

and $[(S.E\ /\ vol)_{Y.P.}]_{T.T} = \frac{1}{2E}(\frac{S_{yt}}{N})^{\wedge 2}$ in equation (Total Strain Energy per unit volume $\leq$ Strain energy per unit volume at yield point under tension test), the following equation is obtained,

$$\sigma_1^{\ 2} + \sigma_2^{\ 2}\sigma_3^{\ 2} - 2\mu(\sigma_1\sigma_2 + \sigma_2\sigma_3 + \sigma_3\sigma_1) \leq (\frac{S_{yt}}{N})^{\wedge 2}$$

for biaxial state of stress, $\sigma_3 = 0$

$$\sigma_1^{\ 2} + \sigma_2^{\ 2} - 2\mu\sigma_1\sigma_2 \leq (\frac{S_{yt}}{N})^{\wedge 2}$$

1. Equaton $\sigma_1^{\ 2} + \sigma_2^{\ 2} - 2\mu\sigma_1\sigma_2 \leq (\frac{S_{yt}}{N})^{\wedge 2}$ is an equation of ellipse ($x^2 + y^2 - xy = a^2$).

2. Semi major axis of the ellipse $= \frac{Syt}{\sqrt{1-\mu}} = \frac{Syt}{\sqrt{0.7}} = 1.2S_{yt}$

   Semi minor axis of the ellipse $= \frac{Syt}{\sqrt{1+\mu}} = \frac{Syt}{\sqrt{1.3}} = 0.87S_{yt}$

   For $\mu = 0.3$

3. Total strain energy theory is suitable under hydrostatic stress condition.

Maximum Distortion Energy Theory (M.D.E.T)

Condition for failure,

Maximum Distortion Energy/ volume (M.D.E/ vol $>$ Distortion energy/ volume at yield point under tension test $(D.E/\ vol)_{Y.P.}]_{T.T}$

Condition for safe design,

Maximum Distortion Energy/ volume $\leq$ Distortion energy/ volume at yield point under tension test

T.S.E/vol = Volumetric S.E/vol + D.E/vol

$$D.E/vol = T.S.E/vol - Volumetric\ S.E/vol$$

Under hydrostatic stress condition, D.E/vol = 0

And

Under pure shear stress condition, Volumetric S.E/vol = 0

From equation $T.S.E./vol = \frac{1}{2E}\left[\sigma_1^2 + \sigma_2^2 + \sigma_3^2 - 2\mu\left(\sigma_1\sigma_2 + \sigma_2\sigma_3 + \sigma_3\sigma_1\right)\right]$

$$T.S.E./vol = \frac{1}{2E}\left[\sigma_1^2 + \sigma_2^2 + \sigma_3^2 - 2\mu\left(\sigma_1\sigma_2 + \sigma_2\sigma_3 + \sigma_3\sigma_1\right)\right]$$

$$\text{Volumetric S.E/ vol} = \frac{1}{2}(\text{Average stress})(\text{Volumetric strain})$$

$$= \frac{1}{2}\left(\frac{\sigma_1 + \sigma_2 + \sigma_3}{3}\right)\left[\left(\frac{1-2\mu}{E}\right)(\sigma_1 + \sigma_2 + \sigma_3)\right]$$

$$\text{Vol S.E/ vol} = \frac{1-2\mu}{6E}(\sigma_1 + \sigma_2 + \sigma_3)^2$$

From equation $D.E/vol = T.S.E/vol - Volumetric\ S.E/vol$ and $\text{Vol S.E/ vol} = \frac{1-2\mu}{6E}(\sigma_1 + \sigma_2 + \sigma_3)^2$

$$D.E/vol = \frac{1+\mu}{6E}\left[(\sigma_1 - \sigma_2)^2 + (\sigma_2 - \sigma_3)^2 + (\sigma_3 - \sigma_1)^2\right]$$

To get $\left[(D.E/vol)_{Y.P.}\right]_{T.T,}$

Substitute $\sigma_1 = \sigma = \frac{Syt}{N}, \sigma_2 = \sigma_3 = 0$ in equation $D.E/vol = \frac{1+\mu}{6E}\left[(\sigma_1 - \sigma_2)^2 + (\sigma_2 - \sigma_3)^2 + (\sigma_3 - \sigma_1)^2\right]$

$$[(D.E/vol)_{Y.P.}]_{T.T} = \frac{1+\mu}{3E}\left(\frac{S_{yt}}{N}\right)^{\wedge 2}$$

Substituting equation $D.E/vol = \frac{1+\mu}{6E}\left[(\sigma_1 - \sigma_2)^2 + (\sigma_2 - \sigma_3)^2 + (\sigma_3 - \sigma_1)^2\right]$

and $[(D.E/vol)_{Y.P.}]_{T.T} = \frac{1+\mu}{3E}\left(\frac{S_{yt}}{N}\right)^{\wedge 2}$ in the condition for safe design, the following equation is obtained

$$[(\sigma_1-\sigma_2)^2+(\sigma_2-\sigma_3)^2+(\sigma_3-\sigma_1)^2]=2(\frac{S_{yt}}{N})^{\wedge 2}$$

For biaxial state of stress, $\sigma_3 = 0$

$$\sigma_1^2+\sigma_2^2-\sigma_1\sigma_2 \le (\frac{S_{yt}}{N})^{\wedge 2}$$

1. Equation $\sigma_1^2+\sigma_2^2-\sigma_1\sigma_2 \le (\frac{S_{yt}}{N})^{\wedge 2}$ is an equation of ellipse.
2. Semi major axis of the ellipse $= \sqrt{2}\,S_{yt}$.

   Semi minor axis of the ellipse $= \sqrt{\frac{2}{3}}\,S_{yt}$.
3. This theory is best theory of failure for ductile material. It gives safe and economic design.
4. This theory is not suitable under hydrostatic stress condition.

Ration of $\frac{S_{YS}}{S_{Yt}}$ by using theories of failure

1. S_{ys} (Yield strength in shear) is obtained from torsion test.
2. Torsion test is conducted under pure torsion i.e. pure shear state of stress

   $(\sigma_x=\sigma_y=0;\ \tau_{xy}=\tau)$.
3. Under pure shear state of stress

   $\sigma_1=\tau, \sigma_2=-\tau$ and $\tau=\frac{16T}{\pi d^3}$
4. S_{ys} can also be obtained by applying theories of failure for pure shear state of stress condition
5. When yielding in shear occurs under pure shear state of stress, $\tau = S_{ys}$

$\frac{S_{ys}}{S_{yt}}$ in Maximum Principal Stress Theory

According to M.P.S.T,

Considering Factor of safety (N) = 1

$\sigma_1 \le S_{yt}$ or

$\sigma_1 = S_{yt}$

But in pure shear state of stress, $\sigma_1 = \tau$

$\tau = S_{yt}$

When yielding occurs in shear under pure shear state of stress, $\tau = S_{ys}$

$$S_{ys} = S_{yt}$$

$$\frac{S_{ys}}{S_{yt}} = 1$$

$\frac{S_{ys}}{S_{yt}}$ in Maximum Shear Stress Theory

According to M.S.S.T,

$$|\sigma_1 - \sigma_2| \le S_{yt}$$

But in pure shear state of stress, $\sigma_1 = \tau$ and $\sigma_2 = -\tau$

$$\tau - (-\tau) = S_{yt}$$

$$2\tau = S_{yt}$$

When yielding occurs in shear under pure shear state of stress, $\tau = S_{ys}$

$$\frac{S_{ys}}{S_{yt}} = \frac{1}{2}$$

$\frac{S_{ys}}{S_{yt}}$ in Maximum Principal Strain Theory

According to M.P.St.T,

$$\sigma_1 - \mu(\sigma_2) = S_{yt}$$

$$\tau - \mu(-\tau) = S_{yt}$$

$$\tau(1+\mu) = S_{yt}$$

$$S_{ys} = \frac{S_{yt}}{1+\mu}$$

for $\mu = 0.3$

$$\frac{S_{ys}}{S_{yt}} = 0.77$$

$\frac{S_{ys}}{S_{yt}}$ in Total Strain Energy Theory

According to T.St.E.T,

$$\sigma_1^2 + \sigma_2^2 - 2\mu\,\sigma_1\sigma_2 = S_{yt}^2$$

$$\tau^2 + \tau^2 + 2\tau^2 = S_{yt}^2$$

$$\tau = \frac{S_{yt}}{\sqrt{2(1+\mu}}$$

$$S_{ys} = \frac{S_{yt}}{\sqrt{2(1+\mu}}$$

for $\mu = 0.3$

$$\frac{S_{ys}}{S_{yt}} = 0.62$$

$\frac{S_{ys}}{S_{yt}}$ in Maximum Distortion Energy Theory

According to M.D.E.T

$$\sigma_1^2 + \sigma_2^2 - \sigma_1\sigma_2 = S_{yt}^2$$

$$\tau^2 + \tau^2 + \tau^2 = S_{yt}^2$$

$$\tau = \frac{S_{yt}}{\sqrt{3}}$$

$$S_{ys} = \frac{S_{yt}}{\sqrt{3}}$$

$$\frac{S_{ys}}{S_{yt}} = 0.577$$

Equivalent Bending Moment (M_e) and Twisting Moment (T_e) Equations

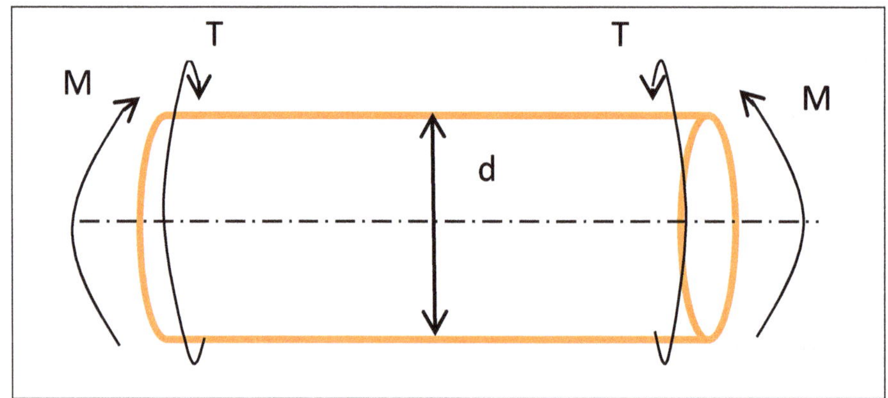

These equations should be used when the component is subjected to both Bending Moment and Twisting Moment simultaneously.

T.O.F	M_e and T_e Equations
M.P.S.T	$M_e = 1\frac{1}{2}\left[M + \sqrt{M^2 + T^2}\right] = \frac{\pi}{32} d^3 \sigma_{per}$
M.S.S.T	$T_e = \sqrt{M^2 + T^2} = \frac{\pi}{16} d^3 \tau_{per}$
M.D.E.T	$M_e = \sqrt{M^2 + \frac{3}{4}T^2} = \frac{\pi}{32} d^3 \sigma_{per}$

Normal Stress Equations (σ_t Equations)

Normal stress equations should be used when a point in a component is subjected to normal stress in one direction only and a shear stress.

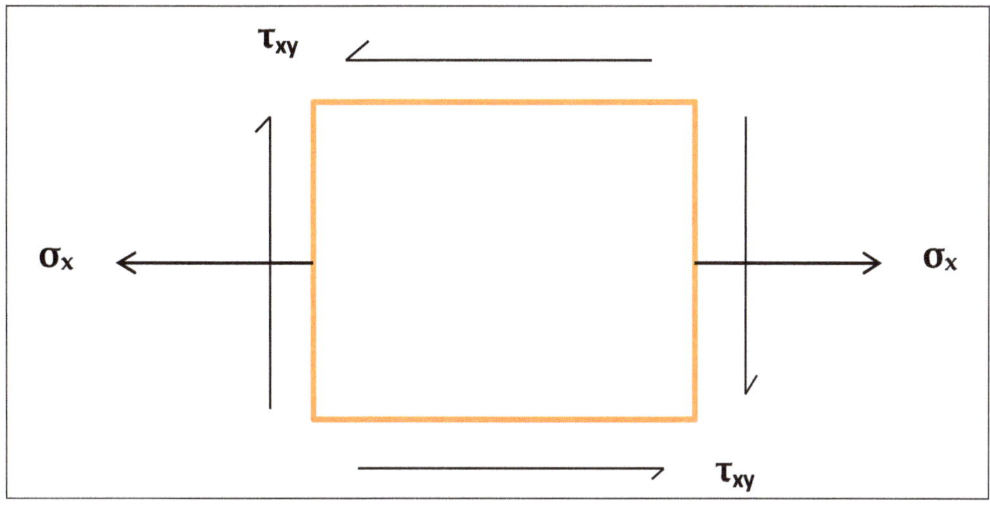

T.O.F	σ_t equations
M.P.S.T	$\sigma_t = \frac{1}{2}\left[\sqrt{\sigma_x^2 + 4\tau_{xt}^2}\right] = \frac{S_{yt}}{N}$
M.S.S.T	$\sigma_t = \sqrt{\sigma_x^2 + 4\tau_{xt}^2} = \frac{S_{yt}}{N}$
M.D.E.T	$\sigma_t = \sqrt{\sigma_x^2 + 3\tau_{xt}^2} = \frac{S_{yt}}{N}$

Shape of Safe Boundaries for Theories of Failure

Graphical representation or safe boundaries are used to check whether the given dimensions of a component are safe or not under given loading conditions.

As per theories of failure for ductile material, S_{yc}=- S_{yt}.

(a) M.P.S.T: Square

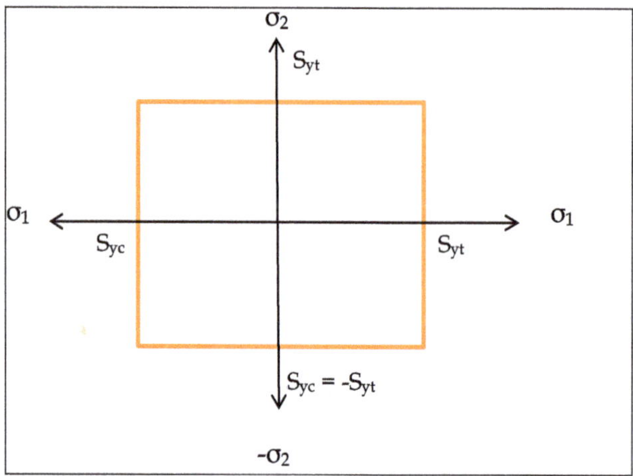

(b) M.S.S.T: Hexagon

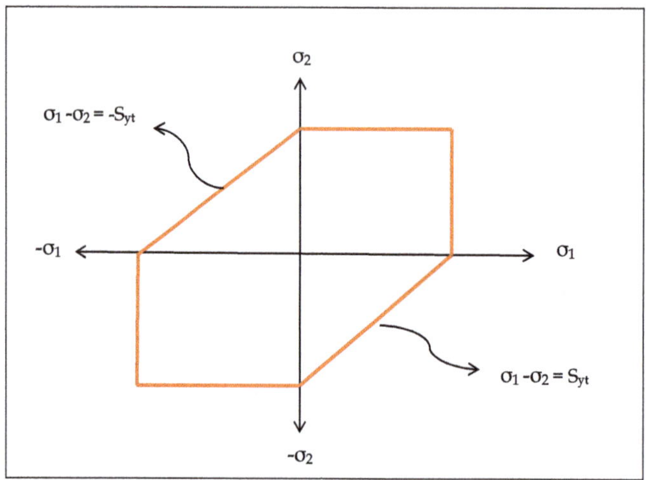

(c) M.P.St.T: Rhombus

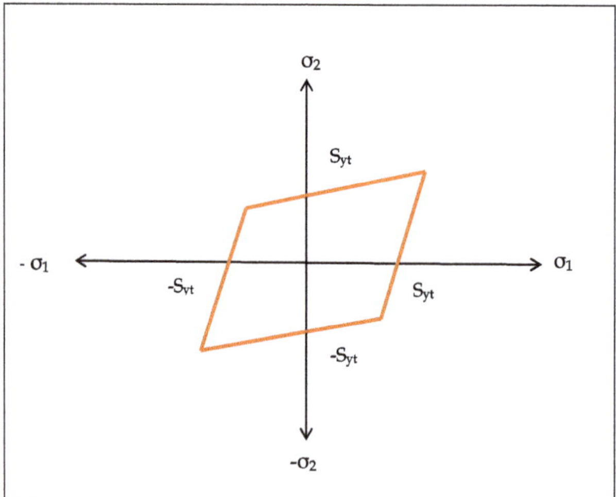

(c) M.D.E.T: Ellipse

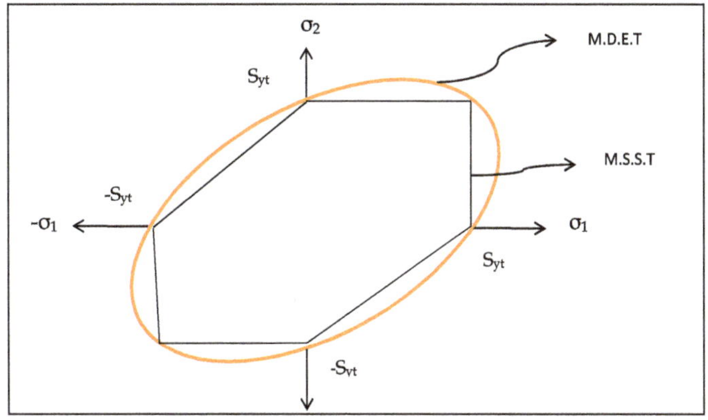

1. Semi major axis of the ellipse = $\sqrt{2}S_{yt}$

   Semi minor axis of the ellipse $= \sqrt{\frac{2}{3}}S_{yt}$

2. As the area bounded by the curve increases, failure stresses increases thereby decreases dimensions and hence cost of safety.

   In all the quadrants

   Area bounded by the MDET curve > Aread bounded by MSST curve

   Hence

   $$(\text{Dimensions})_{\text{MDET}} > (\text{Dimensions})_{\text{MSST}}$$

(c) T.St.E.T: Ellipse

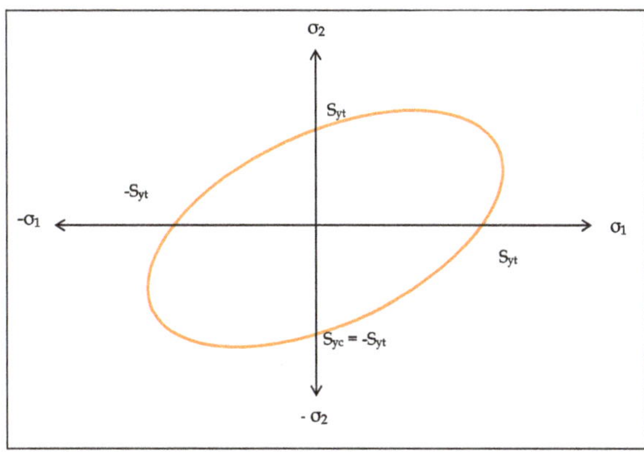

Semi- major axis of the ellipse $= \frac{S_{yt}}{\sqrt{1-\mu}}$

Semi- minor axis of the ellipse $= \frac{S_{yt}}{\sqrt{1+\mu}}$.

References

- Strain-mechanics, science: britannica.com, Retrieved 17 April, 2019
- Spear and Dismukes (1994). Synthetic Diamond – Emerging CVD Science and Technology. Wiley, N.Y. p. 315. ISBN 978-0-471-53589-8
- Barber, A. H.; Lu, D.; Pugno, N. M. (2015). "Extreme strength observed in limpet teeth". Journal of the Royal Society Interface. 12 (105): 105. doi:10.1098/rsif.2014.1326. PMC 4387522
- Theories-of-failure: thegateacademy.com, Retrieved 10 June, 2019
- Bodros, E. (2002). "Analysis of the flax fibres tensile behaviour and analysis of the tensile stiffness increase". Composite Part A. 33 (7): 939–948. doi:10.1016/S1359-835X(02)00040-4

6

Tools and Methods in Mechanical Engineering

There are a wide variety of tools used within mechanical engineering such as computer-aided design, computer –aided manufacturing and computational fluid dynamics. The use of computers to assist in creating, modifying, analyzing and optimizing a design is known as computer-aided design. The diverse applications of these tools in the field of mechanical engineering have been thoroughly discussed in this chapter.

COMPUTER-AIDED ENGINEERING

CAE or Computer-Aided Engineering is a term used to describe the procedure of the entire product engineering process, from design and virtual testing with sophisticated analytical algorithms to the planning of manufacturing. Computer-aided engineering is standard in almost any industry that uses some sort of design software to develop products. CAE is the next step in not only designing a product, but also supporting the engineering process, as it allows performing tests and simulations of the product's physical properties without needing a physical prototype. In the context of CAE, the most commonly used simulation analysis types include Finite Element Analysis, Computational Fluid Dynamics, Thermal Analysis, Multibody Dynamics and Optimizations.

By leveraging the advantages of engineering simulation, especially when combined with the power and the speed of high-performance cloud computing, the cost and time of each design iteration cycle as well as the overall development process can be considerably reduced. The standard CAE workflow is to first generate an initial design and then simulate the CAD geometry. The simulation results are then evaluated and used to improve the design. This process is repeated until all the product's requirements are met and virtually confirmed. In case of any weak spots or areas where the digital prototype's performance doesn't match the expectations, engineers and designers can improve the CAD model and check the effects of their change by testing the updated design in a new simulation. This process supports faster product development as there is no need for building physical prototypes in early development stages. Simulating with CAE methods will only take a few hours at most, in comparison to days or probably weeks that building a physical prototype would require. Everyone familiar with the product development process knows that it is inevitable to build a physical prototype before starting with the serial production of a product, but simulation can help to reduce the amount of those prototypes. When planning to integrate simulation techniques into the product development process, it is important to know about the environment (forces, temperatures, etc.) that the product is going to be exposed to. Knowing these conditions is crucial to properly set up a simulation. The predictive value of any simulation can only be the precision of the boundary conditions made. Up until now, besides predicting the environmental factors and conditions, engineering simulation was a complex and difficult endeavor by itself, mostly reserved for experienced engineers and simulation experts. Beginners had to struggle with a steep learning curve. Modern CAE simulation tools, such as SimScale, try to break down these barriers, allowing even inexperienced users without deep knowledge of the physical processes and special solver characteristics to produce insightful simulation results.

Simulating complex geometries is very difficult, even for modern computers. This is why a lot of computing power is needed to perform realistic simulation results. Big companies with sophisticated IT infrastructure can use their own servers to host and run simulations. The rise of HPC cloud computing now also gives smaller companies, which usually can't afford to buy and maintain the necessary hardware, access to the same simulation tools and capabilities that previously were reserved only for a select few. This disruption in the market for simulation products makes it now possible for everyone to simulate the products they design.

Fields of Application

CAE can be used in almost any industry and company that designs a product exposed to different environments. Industries using computer-aided engineering in their product development process include but are not limited to: automotive, aerospace, plant engineering, electronics, energy, consumer goods and HVAC. The products that can be simulated range from extremely small parts of products to very big and complex structures such as race cars, bridges or even power plants.

Testing the structural integrity of a crane that carries a specific load to a rooftop is a possible application as well as assessing the acoustic design of a concert hall or the convective flow inside a light bulb; all these are examples of applications where simulation can make a huge, sometimes life-saving difference.

COMPUTER-AIDED DESIGN

CAD stands for Computer Aided Design (and drafting, depending on the industry) and is computer software used to create 2D and 3D models and designs.

CAD software is used across many different industries and occupations, and can be used to make architectural designs, building plans, floor plans, electrical schematics, mechanical drawings, technical drawings, blueprints and even the special effects in your favorite movies and TV shows.

Benefits of CAD

Prior to the advent of computer aided design, designs needed to be manually drawn using pencil and paper. Every object, line or curve needed to be drawn by hand using rulers, protractors and other drafting tools. Calculations, such as the structural load on a building component, would need to be done manually by an engineer or designer, a very time consuming - and error prone - process.

CAD software changed all of this. Designs can be created and edited in much less time, as well as saved for future use. CAD drawings are not limited to the 2D space of a piece of paper, and can be viewed from many different angles to ensure proper fit and design. Calculations are performed by the computer, making it much easier to test the viability of designs. Designs can be shared and collaborated on in real time, greatly decreasing the overall time needed to complete a drawing.

Types of CAD Drawings

There are a wide variety of uses for CAD software and the types of designs that can be made. Below are some common designs and drawings that can be made with CAD software.

Floor Plans

Floor plans are scaled diagrams that show the size, placement and shape of rooms and other objects within a structure using a top down view. Floor plans help to visualize the footprint of a building, home or other structure. Floor plans are great for laying out objects, like furniture, within a structure to ensure a proper fit.

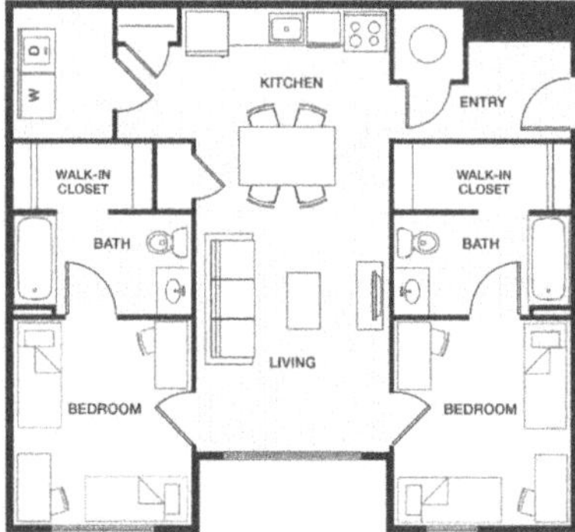

Technical Drawings and Blueprints

A technical drawing is a detailed, scaled plan or drawing of an object. Technical drawings are used to deliver exact specifications of how something should be made. Technical drawings can include architectural, mechanical and engineering designs. Blueprints are reproductions of technical drawings, but the word blueprint is also used to describe any type of plan, such as a floor plan.

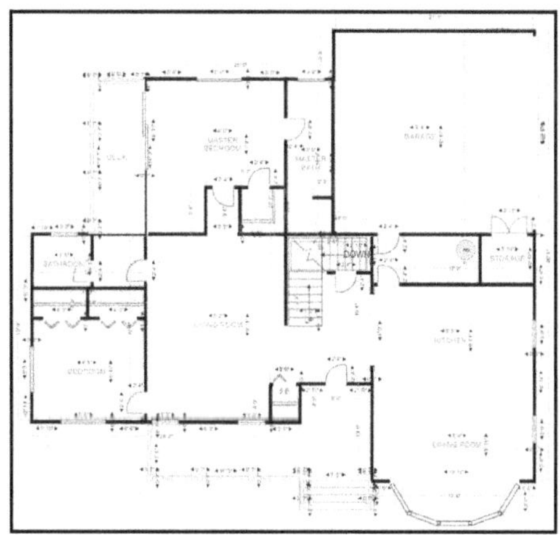

Piping and Instrumentation Diagrams

A piping & instrumentation diagram (P&ID) shows the relationships between piping, instrumentation and other system components in a physical process flow. For example, a P&ID can show the types of valves, pumps, tanks and other components within the larger system, and how they connect to, and interact with, one another.

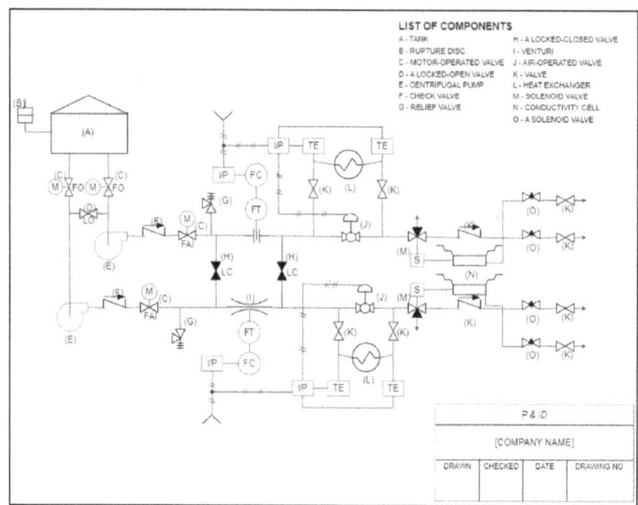

HVAC Diagrams

Heating, ventilation, and air conditioning (HVAC) drawings provide information about the ventilation, heating and air conditioning systems within a given location. They can include the size and location of ductwork, connections to control units, as well as the relationship and connections between various components.

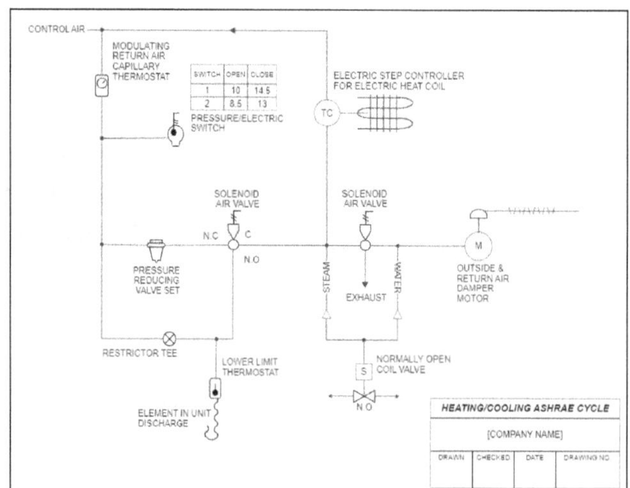

Site and Plot Plans

Site plans, also known as plot plans, are top down view, scaled drawings showing the proposed usage and development of a piece of land. Site plans can include the footprint of buildings, landscaping designs, walkways, parking lots, drainage and water lines, and will show the placement of all of these items relative to one another.

Electrical Schematics

Electrical schematics provide an overview of what components are included in an electrical system, and the relationship between those components. Electrical schematics typically use symbols to represent the various components and elements within an electrical system. For more granularity regarding placement of the electrical components, and how wires connect to them and each other, a wiring diagram would be more useful.

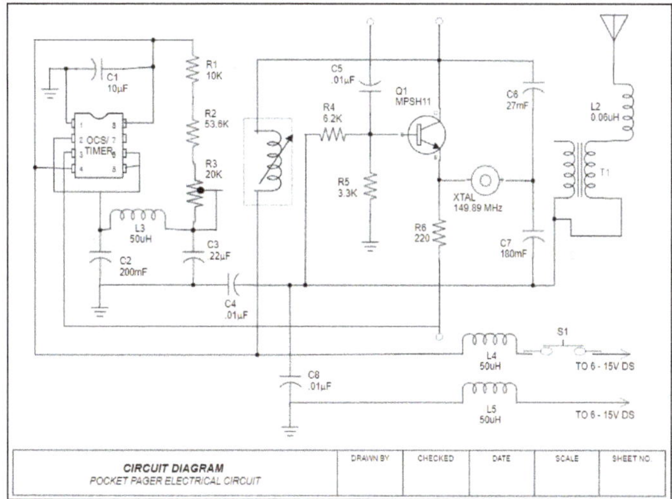

Wiring Diagrams

Wiring diagrams show the actual connection of wires to each other and to other components in an electrical system, as well as where the components are physically located within the system. Unlike electrical schematics, which provide a broad overview of the components in an electrical system and their relationship to one another, wiring diagrams show where wires actually connect to one another, and to the other components. They also show where the components will be located relative to one another.

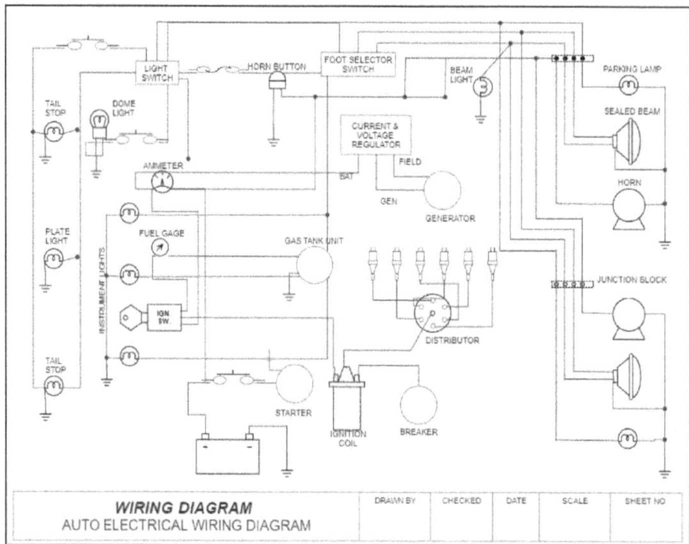

COMPUTER-AIDED MANUFACTURING

Computer Aided Manufacturing (CAM) is the use of software and computer-controlled machinery to automate a manufacturing process. Based on that definition, you need three components for a CAM system to function:

- Software that tells a machine how to make a product by generating toolpaths.
- Machinery that can turn raw material into a finished product.
- Post Processing that converts toolpaths into a language machines can understand.

These three components are glued together with tons of human labor and skill. As an industry we've spent years building and refining the best manufacturing machinery around. Today, there's no design too tough for any capable machinist shop to handle.

CAD to CAM Process

Without CAM, there is no CAD. CAD focuses on the design of a product or part. How it looks, how it functions. CAM focuses on how to make it. You can design the most elegant part in your CAD tool, but if you can't efficiently make it with a CAM system then you're better off kicking rocks.

The start of every engineering process begins in the world of CAD. Engineers will make either a 2D or 3D drawing, whether that's a crankshaft for an automobile, the inner skeleton of a kitchen faucet, or the hidden electronics in a circuit board. In the world of CAD, any design is called a model and contains a set of physical properties that will be used by a CAM system.

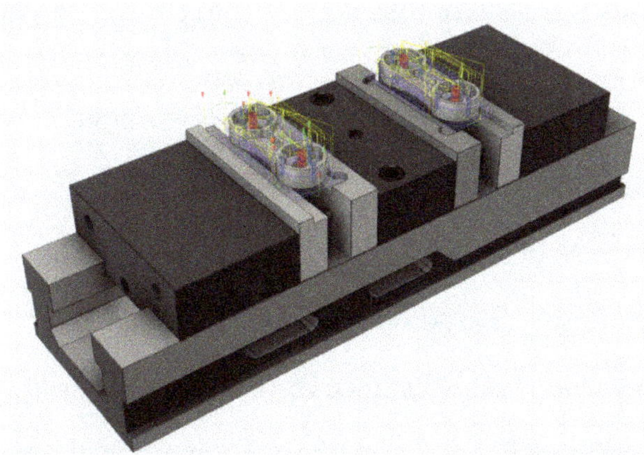

When a design is complete in CAD, it can then be loaded into CAM. This is traditionally done by exporting a CAD file and then importing it into CAM software. If you're using a tool like Fusion 360, both CAD and CAM exist in the same world, so there's no import/export required.

Once your CAD model is imported into CAM, the software starts preparing the model for machining. Machining is the controlled process of transforming raw material into a defined shape through actions like cutting, drilling, or boring.

CAM software prepares a model for machining by working through several actions, including:

- Checking if the model has any geometry errors that will impact the manufacturing process.
- Creating a toolpath for the model, which is a set of coordinates the machine will follow during the machining process.
- Setting any required machine parameters including cutting speed, voltage, cut/pierce height, etc.
- Configuring nesting where the CAM system will decide the best orientation for a part to maximize machining efficiency.

Running a Contour toolpath in Fusion 360.

Once the model is prepared for machining, all of that information gets sent to a machine to physically produce the part. However, we can't just give a machine a bunch of instructions in English, we need to speak the machine's language. To do this we convert all of our machining information to a language called G-code. This is the set of instructions that controls a machine's actions including speed, feed rate, coolants, etc.

G-code is easy to read once you understand the format. An example looks like this:

G01 X1 Y1 F20 T01 S500

This breaks down from left to right as:

- G01 indicates a linear move, based on coordinates X1 and Y1.
- F20 sets a feed rate, which is the distance the machine travels in one spindle revolution.
- T01 tells the machine to use Tool 1, and S500 sets the spindle speed.

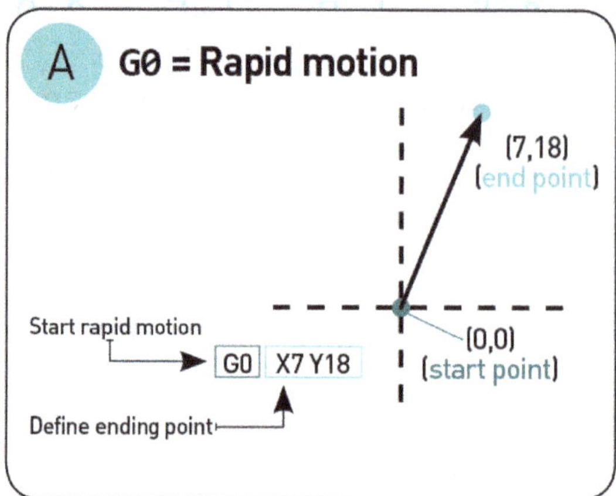

Once the G-code is loaded into the machine and an operator hits start, our job is done. Now it's time to let the machine do the job of executing G-code to transform a raw material block into a finished product.

CNC Machines at a Glance

Up until this point we've talked about the machinery in a CAM system as simply machines, but that really doesn't do them justice. Watching a Haas milling machine slide through a block of metal like it's butter puts a smile on our face.

All modern manufacturing centers will be running a variety of Computer Numerical Control (CNC) machines to produce engineered parts. The process of programming a CNC machine to perform specific actions is called CNC machining.

Before CNC machines came to be, manufacturing centers were operated manually by Machinist veterans. Of course, like all things that computers touch, automation soon followed. These days the only human intervention required for running a CNC machine is loading a program, inserting raw material, and then unloading a finished product.

CNC Routers

These machines cut parts and carve out a variety of shapes with high speed spinning components. For example, a CNC router used for woodworking can make easy work of cutting plywood into cabinet parts. It can also easily tackle complex decorative engraving on a door panel. CNC routers have 3-axis cutting capabilities, which allow them to move along the X, Y, and Z axes.

Water, Plasma and Laser Cutters

These machines use precise lasers, high pressure water, or a plasma torch to perform a controlled cut or engraved finished. Manual engraving techniques can take months to complete by hand, but one of these machines can complete the same work in hours or days. Plasma cutters are especially useful for cutting through electrically conductive materials like metals.

Milling Machines

These machines chip away at a variety of materials like metal, wood, composites, etc. Milling machines have enormous versatility with a variety of tools that can accomplish specific material and shape requirements. The overall goal of a milling machine is to remove mass from a raw block of material as efficiently as possible.

Lathes

These machines also chip away at raw materials like a milling machine, they just do it differently. A milling machine has a spinning tool and stationary material, where a lathe spins the material and cuts with a stationary tool.

Electrical Discharge Machines (EDM)

These machines cut a desired shape out of raw material through an electrical discharge. An electrical spark is created between an electrode and raw material, with the spark's temperature reaching 8,000 to 12,000 degrees Celsius. This allows an EDM to melt through nearly anything in a controlled and ultra precise process.

The Human Element of CAM

The human element has always been a touchy subject since CAM arrived on scene in the 1990s. Back in the 1950s when CNC machining was first introduced by John T. Parsons, skillfully operating machines required an enormous amount of training and practice.

In the days of manual machining, being a Machinist was a badge of honor that took years of training to perfect. A Machinist had to do it all – read blueprints, know which tools to use, define feeds and speeds for specific materials, and carefully cut a part by hand. It wasn't just about precise manual dexterity. Being a Machinist was, and still is, both an art and a science.

These days, the modern machinist is alive and well as man, machine, and software combine to move our industry forward. Skills that used to take 40 years to master can now be conquered in a fraction of the time. New machines and CAM software have given us more control than ever to design and make better and more innovative products than our forefathers, which they'll admit begrudgingly.

What does all this mean for the human element of manufacturing? The role of a Traditional Machinist is shifting. Today we're seeing an environment of Modern Machinists played out with three typical roles:

- The Operator: This individual loads raw materials into a CNC machine and run completed parts through the final packaging process.
- The Setup Operator: This individual performs the initial configuration for a CNC machine, which includes loading a G-code program and setting up tools.
- The Programmer: This individual takes the drawing for a CAD model and decides how to make it with their available CNC machines. Their job is to define the toolpaths, tools, speeds, and feeds in the G-code to get the job done.

In a typical workflow the Programmer will hand off his program to the Setup Operator, who will then load the G-code into the machine. Once the machine is ready to roll, the Operator will then make the part. In some shops these roles might combine and overlap into the responsibilities of one or two people.

Outside of day-to-day machine operations, there is also the Manufacturing Engineer on staff. In a new shop setup, this individual typically establishes systems and determines an ideal manufacturing process. For existing setups, a Manufacturing Engineer will monitor equipment and product quality while handling other managerial tasks.

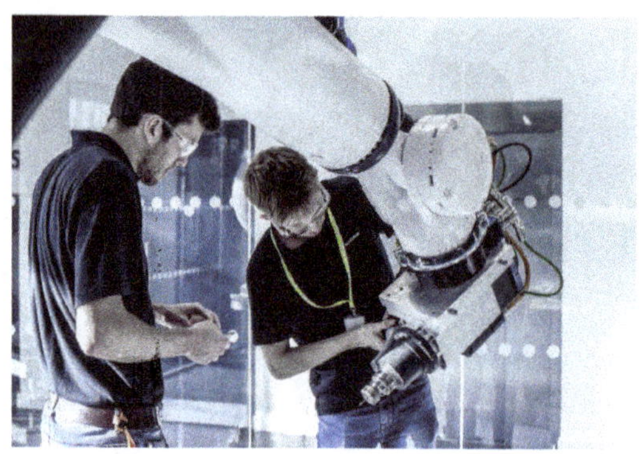

The Impact of CAM

We have John T. Parsons to thank for introducing a punch card method to program and automate machinery. In 1949 the United States Air Force funded Parsons to build an automated machine that could outperform manual NC machines. With some help from MIT, Parsons was able to develop the first NC prototype.

From there the world of CNC machining started to take off. In the 1950s the United States Army bought NC machine and loaned them out to manufacturers. The idea was to incentive companies to adopt the new technology into their manufacturing process. During this time we also saw MIT develop the first universal programming language for CNC machines: G-code.

Code	Group	Description	Modal	Page
G00	1	Rapid move	Y	10
G01	1	Linear feed move	Y	10
G02	1	Clockwise Arc feed move	Y	11
G03	1	Counter clockwise Arc feed Move	Y	11
G04	0	Dwell	N	14
G09	0	Exact stop	N	14
G10	0	Fixture and tool offset setting	N	15
G12	1	Clockwise circle	Y	18
G13	1	Counter clockwise circle	Y	18
G15	11	Polar coordinate cancel	Y	18
G16	11	Polar coordinate	Y	18
G17	2	XY plane select	Y	20
G18	2	ZX Plane select	Y	20
G19	2	YZ plane select	Y	20
G20	6	Inch	Y	20

G21	6	Millimeter	Y	20
G28	0	Zero Return	N	21
G30	0	2nd,3rd,4^{th}Zero Return	N	22
G31	1	Probe Function	N	22
G32	1	Threading*	N	23
G40	7	Cutter Compensation Cancel	Y	23
G41	7	Cutter Compensation left	Y	25
G42	7	Cutter Compensation right	Y	25
G43	8	Tool length Offset + Enable	Y	25
G44	8	Tool length Offset- Enable	Y	25
G49	8	Tool length Offset cancel	`y	25
G50	9	Cancel scaling	Y	25
G51	9	Scale Axes	Y	25
G52	0	Local coordinate System Shift	Y	26

The 1990s brought the introduction of CAD and CAM on the PC, and has completely revolutionized how we approach manufacturing today. The earliest CAD and CAM jobs were reserved for expensive automotive and aerospace applications, but today software like Fusion 360 is available for manufacturing shops of any shape and size. Since its inception, CAM has delivered a ton of improvements to the manufacturing process, including:

- Improved machine capabilities: CAM systems can take advantage of advanced 5-axis machinery to deliver more sophisticated and higher quality parts.
- Improved machine efficiency: Today's CAM software provides high-speed machine tool paths that help us manufacture parts faster than ever.
- Improved material usage: With additive machinery and CAM systems, we're able to produce complex geometries with minimal waste which means lower costs.

Of course, these benefits have some trade-offs. CAM systems and machinery require a massive upfront cost. For example, a Haas VF-1 costs about $45k out the door; now imagine an entire shop floor of those. There's also the problem of turnover. With machine operation becoming less of a skilled trade, it's incredibly hard to attract and retain good talent.

FINITE ELEMENT ANALYSIS

The finite element method (FEM) is a numerical method for solving problems of engineering and mathematical physics. Typical problem areas of interest include structural analysis, heat transfer, fluid flow, mass transport, and electromagnetic potential. The analytical solution of these problems generally require the solution to boundary value problems for partial differential equations. The finite element method formulation of the problem results in a system of algebraic equations. The method approximates the unknown function over the domain. To solve the problem, it subdivides a large system into smaller, simpler parts that are called finite elements. The simple equations that model these finite elements are then assembled into a larger system of equations that models the entire problem. FEM then uses variational methods from the calculus of variations to approximate a solution by minimizing an associated error function.

Studying or analyzing a phenomenon with FEM is often referred to as finite element analysis (FEA).

Basic Concepts

The subdivision of a whole domain into simpler parts has several advantages:

- Accurate representation of complex geometry.
- Inclusion of dissimilar material properties.
- Easy representation of the total solution.
- Capture of local effects.

A typical work out of the method involves dividing the domain of the problem into a collection of subdomains, with each subdomain represented by a set of element equations to the original problem, followed by systematically recombining all sets of element equations into a global system of equations for the final calculation. The global system of equations has known solution techniques, and can be calculated from the initial values of the original problem to obtain a numerical answer.

In the first step above, the element equations are simple equations that locally approximate the original complex equations to be studied, where the original equations are often partial differential equations (PDE). To explain the approximation in this process, FEM is commonly introduced as a special case of Galerkin method. The process, in mathematical language, is to construct an integral of the inner product of the residual and the weight functions and set the integral to zero. In simple terms, it is a procedure that minimizes the error of approximation by fitting trial functions into the PDE. The residual is the error caused by the trial functions, and the weight functions are polynomial approximation functions that project the residual. The process eliminates all the spatial derivatives from the PDE, thus approximating the PDE locally with:

- A set of algebraic equations for steady state problems,
- A set of ordinary differential equations for transient problems.

These equation sets are the element equations. They are linear if the underlying PDE is linear, and vice versa. Algebraic equation sets that arise in the steady state problems are solved using numerical linear algebra methods, while ordinary differential equation sets that arise in the transient problems are solved by numerical integration using standard techniques such as Euler's method or the Runge-Kutta method.

A global system of equations is generated from the element equations through a transformation of coordinates from the subdomains' local nodes to the domain's global nodes. This spatial transformation includes appropriate orientation adjustments as applied in relation to the reference coordinate system. The process is often carried out by FEM software using coordinate data generated from the subdomains.

FEM is best understood from its practical application, known as finite element analysis (FEA). FEA as applied in engineering is a computational tool for performing engineering analysis. It includes the use of mesh generation techniques for dividing a complex problem into small elements, as well as the use of software program coded with FEM algorithm. In applying FEA, the complex problem is usually a physical system with the underlying physics such as the Euler-Bernoulli beam equation, the heat equation, or the Navier-Stokes equations expressed in either PDE or integral equations, while the divided small elements of the complex problem represent different areas in the physical system.

FEA is a good choice for analyzing problems over complicated domains (like cars and oil pipelines), when the domain changes (as during a solid state reaction with a moving boundary), when the desired precision varies over the entire domain, or when the solution lacks smoothness. FEA simulations provide a valuable resource as they remove multiple instances of creation and testing of hard prototypes for various high fidelity situations. For instance, in a frontal crash simulation it is possible to increase prediction accuracy in "important" areas like the front of the car and reduce it in its rear (thus reducing cost of the simulation). Another example would be in numerical weather prediction, where it is more important to have accurate predictions over developing highly nonlinear phenomena (such as tropical cyclones in the atmosphere, or eddies in the ocean) rather than relatively calm areas.

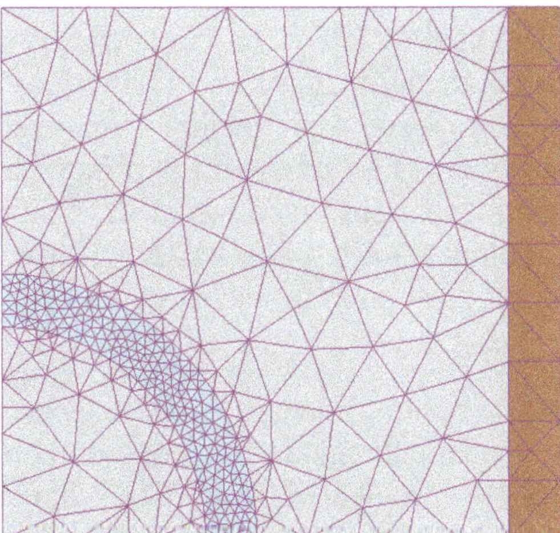

FEM mesh created by an analyst prior to finding a solution to a magnetic problem using FEM software.

Colours indicate that the analyst has set material properties for each zone, in this case a conducting wire coil in orange; a ferromagnetic component (perhaps iron) in light blue; and air in grey. Although the geometry may seem simple, it would be very challenging to calculate the magnetic field for this setup without FEM software, using equations alone.

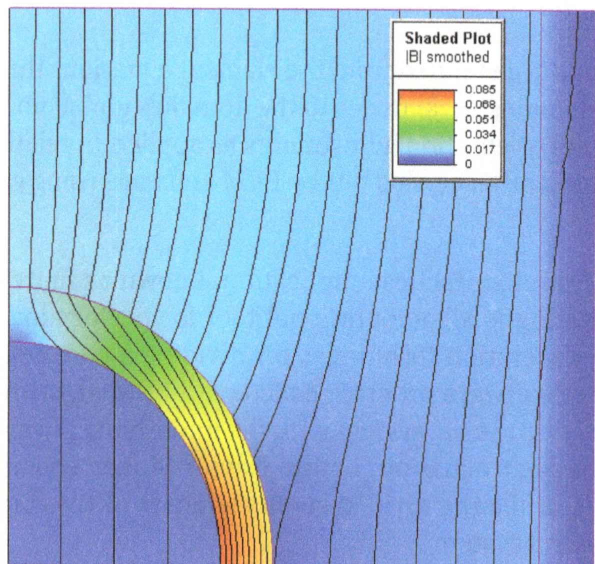

FEM solution to the problem at left, involving a cylindrically shaped magnetic shield.

The ferromagnetic cylindrical part is shielding the area inside the cylinder by diverting the magnetic field created by the coil. The color represents the amplitude of the magnetic flux density, as indicated by the scale in the inset legend, red being high amplitude. The area inside the cylinder is low amplitude (dark blue, with widely spaced lines of magnetic flux), which suggests that the shield is performing as it was designed to.

While it is difficult to quote a date of the invention of the finite element method, the method originated from the need to solve complex elasticity and structural analysis problems in civil and aeronautical engineering. Its development can be traced back to the work by A. Hrennikoff and R. Courant in the early 1940s. Another pioneer was Ioannis Argyris. In the USSR, the introduction of the practical application of the method is usually connected with name of Leonard Oganesyan. In China, in the later 1950s and early 1960s, based on the computations of dam constructions, K. Feng proposed a systematic numerical method for solving partial differential equations. The method was called the finite difference method based on variation principle, which was another independent invention of the finite element method. Although the approaches used by these pioneers are different, they share one essential characteristic: mesh discretization of a continuous domain into a set of discrete sub-domains, usually called elements.

Hrennikoff's work discretizes the domain by using a lattice analogy, while Courant's approach divides the domain into finite triangular subregions to solve second order elliptic partial differential equations (PDEs) that arise from the problem of torsion of a cylinder. Courant's contribution was evolutionary, drawing on a large body of earlier results for PDEs developed by Rayleigh, Ritz, and Galerkin.

The finite element method obtained its real impetus in the 1960s and 1970s by the developments of J. H. Argyris with co-workers at the University of Stuttgart, R. W. Clough with co-workers at UC Berkeley, O. C. Zienkiewicz with co-workers Ernest Hinton, Bruce Irons and others at the Swansea University, Philippe G. Ciarlet at the University of Paris 6 and Richard Gallagher with co-workers at Cornell University. Further impetus was provided in these years by available open source finite element software programs. NASA sponsored the original version of NASTRAN, and UC Berkeley made the finite element program SAP IV widely available. In Norway the ship classification society Det Norske Veritas (now DNV GL) developed Sesam in 1969 for use in analysis of ships. A rigorous mathematical basis to the finite element method was provided in 1973 with the publication by Strang and Fix. The method has since been generalized for the numerical modeling of physical systems in a wide variety of engineering disciplines, e.g., electromagnetism, heat transfer, and fluid dynamics.

Technical Discussion

The Structure of Finite Element Methods

A finite element method is characterized by a variational formulation, a discretization strategy, one or more solution algorithms and post-processing procedures.

Examples of variational formulation are the Galerkin method, the discontinuous Galerkin method, mixed methods, etc.

A discretization strategy is understood to mean a clearly defined set of procedures that cover (a) the creation of finite element meshes, (b) the definition of basis function on reference elements (also called shape functions) and (c) the mapping of reference elements onto the elements of the mesh. Examples of discretization strategies are the h-version, p-version, hp-version, x-FEM, isogeometric analysis, etc. Each discretization strategy has certain advantages and disadvantages. A reasonable criterion in selecting a discretization strategy is to realize nearly optimal performance for the broadest set of mathematical models in a particular model class.

There are various numerical solution algorithms that can be classified into two broad categories; direct and iterative solvers. These algorithms are designed to exploit the sparsity of matrices that depend on the choices of variational formulation and discretization strategy.

Postprocessing procedures are designed for the extraction of the data of interest from a finite element solution. In order to meet the requirements of solution verification, postprocessors need to provide for *a posteriori* error estimation in terms of the quantities of interest. When the errors of approximation are larger than what is considered acceptable then the discretization has to be changed either by an automated adaptive process or by action of the analyst. There are some very efficient postprocessors that provide for the realization of superconvergence.

Illustrative Problems P1 and P2

We will demonstrate the finite element method using two sample problems from which the general method can be extrapolated. It is assumed that the reader is familiar with calculus and linear algebra.

P1 is a one-dimensional problem:

$$\text{P1 }:\begin{cases} u''(x)=f(x) \text{ in } (0,1), \\ u(0)=u(1)=0, \end{cases}$$

where f is given, x is an unknown function of u , and x is the second derivative of u with respect to x.

P2 is a two-dimensional problem (Dirichlet problem):

$$\text{P2 }:\begin{cases} u_{xx}(x,y)+u_{yy}(x,y)=f(x,y) & \text{in } \Omega, \\ u=0 & \text{on } \partial\Omega, \end{cases}$$

Where Ω is a connected open region in the (x,y) plane whose boundary $\partial\Omega$ is nice (e.g., a smooth manifold or a polygon), and u_{xx} and u_{yy} denote the second derivatives with respect to x and y, respectively.

The problem P1 can be solved directly by computing antiderivatives. However, this method of solving the boundary value problem (BVP) works only when there is one spatial dimension and does not generalize to higher-dimensional problems or to problems like $u+u''=f$. For this reason, we will develop the finite element method for P1 and outline its generalization to P2.

Our explanation will proceed in two steps, which mirror two essential steps one must take to solve a boundary value problem (BVP) using the FEM.

- In the first step, one rephrases the original BVP in its weak form. Little to no computation is usually required for this step. The transformation is done by hand on paper.
- The second step is the discretization, where the weak form is discretized in a finite-dimensional space.

After this second step, we have concrete formulae for a large but finite-dimensional linear problem whose solution will approximately solve the original BVP. This finite-dimensional problem is then implemented on a computer.

Weak Formulation

The first step is to convert P1 and P2 into their equivalent weak formulations.

The Weak Form of P1

If u solves P1, then for any smooth function v that satisfies the displacement boundary conditions, i.e. $v=0$ at $x=0$ and $x=1$ we have,

$$\int_0^1 f(x)v(x)\,dx=\int_0^1 u''(x)v(x)\,dx.$$

Conversely, if u with $u(0)=u(1)=0$ satisfies for every smooth function $v(x)$ then one may show that this u will solve P1. The proof is easier for twice continuously differentiable (mean value theorem), but may be proved in a distributional sense as well. We define a new operator or map $\phi(u,v)$ by using integration by parts on the right-hand-side,

$$\begin{aligned}\int_0^1 f(x)v(x)\,dx &= \int_0^1 u''(x)v(x)\,dx \\ &= u'(x)v(x)\Big|_0^1 - \int_0^1 u'(x)v'(x)\,dx \\ &= -\int_0^1 u'(x)v'(x)\,dx \equiv -\phi(u,v),\end{aligned}$$

where we have used the assumption that $v(0) = v(1) = 0$.

The Weak Form of P2

If we integrate by parts using a form of Green's identities, we see that if u solves P2, then we may define $\phi(u,v)$, for any v by:

$$\int_\Omega fv\,ds = -\int_\Omega \nabla u \cdot \nabla v\,ds \equiv -\phi(u,v),$$

where ∇ denotes the gradient and $\cdot$ denotes the dot product in the two-dimensional plane. Once more ϕ can be turned into an inner product on a suitable space $H_0^1(\Omega)$ of once differentiable functions of Ω that are zero on $\partial\Omega$. We have also assumed that $v \in H_0^1(\Omega)$. Existence and uniqueness of the solution can also be shown.

A Proof Outline of Existence and Uniqueness of the Solution

We can loosely think of $H_0^1(0,1)$ to be the absolutely continuous functions of $(0,1)$ that are 0 at $x = 0$ and $x = 1$ such functions are (weakly) once differentiable and it turns out that the symmetric bilinear map ϕ then defines an inner product which turns $H_0^1(0,1)$ into a Hilbert space. On the other hand, the left-hand-side $\int_0^1 f(x)v(x)dx$ is also an inner product, this time on the Lp space $L^2(0,1)$. An application of the Riesz representation theorem for Hilbert spaces shows that there is a unique u solving $= -\int_0^1 u'(x)v'(x)\,dx \equiv -\phi(u,v)$, and therefore P1. This solution is a-priori only a member of $H_0^1(0,1)$, but using elliptic regularity, will be smooth if f is.

Discretization

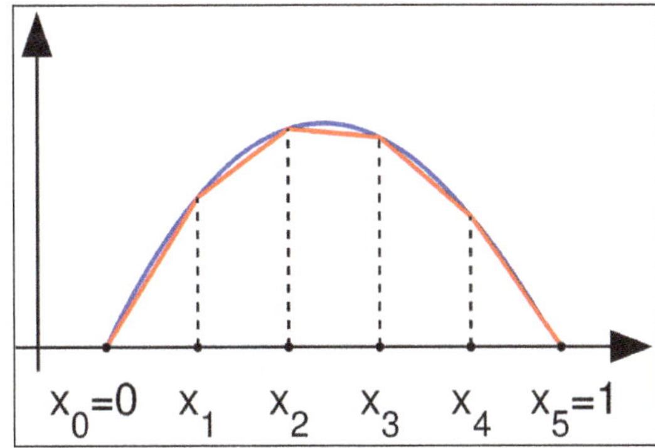

A function in H_0^1 with zero values at the endpoints (blue), and a piecewise linear approximation (red).

P1 and P2 are ready to be discretized which leads to a common sub-problem. The basic idea is to replace the infinite-dimensional linear problem:

Find $u \in H_0^1$ such that,

$$\forall v \in H_0^1, -\phi(u,v) = \int fv$$

with a finite-dimensional version:

Find $u \in V$ such that,

$$\forall v \in V, -\phi(u,v) = \int fv$$

where V is a finite-dimensional subspace of H_0^1. There are many possible choices for V (one possibility leads to the spectral method). However, for the finite element method we take V to be a space of piecewise polynomial functions.

For Problem P1

We take the interval $(0,1)$, choose n values of x with $0 = x_0 < x_1 < \cdots < x_n < x_{n+1} = 1$ and we define V by:

$$V = \{v:[0,1] \to \mathbb{R} : v \text{ is continuous, } v|_{[x_k, x_{k+1}]} \text{ is linear for } k = 0,\ldots,n, \text{ and } v(0) = v(1) = 0\}$$

where we define $x_0 = 0$ and $x_{n+1} = 1$. Observe that functions in V are not differentiable according to the elementary definition of calculus. Indeed, if $v \in V$ then the derivative is typically not defined at any $x = x_k$, $k = 1,\ldots,n$. However, the derivative exists at every other value of x and one can use this derivative for the purpose of integration by parts.

For Problem P2

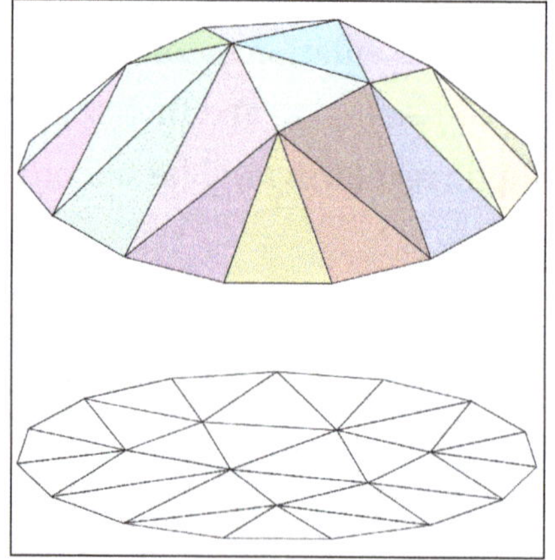

A piecewise linear function in two dimensions.

We need V to be a set of functions of Ω. In the figure on the right, we have illustrated a triangulation of a 15 sided polygonal region Ω in the plane (below), and a piecewise linear function (above, in color) of this polygon which is linear on each triangle of the triangulation; the space V would consist of functions that are linear on each triangle of the chosen triangulation.

One hopes that as the underlying triangular mesh becomes finer and finer, the solution of the discrete problem $\forall v \in V, -\phi(u,v) = \int fv$, will in some sense converge to the solution of the original boundary value problem P2. To measure this mesh fineness, the triangulation is indexed by a real valued parameter $h > 0$ which one takes to be very small. This parameter will be related to the size of the largest or average triangle in the triangulation. As we refine the triangulation, the space of piecewise linear functions V must also change with h. For this reason, one often reads instead of V_h in the literature. Since we do not perform such an analysis, we will not use this notation.

Choosing a Basis

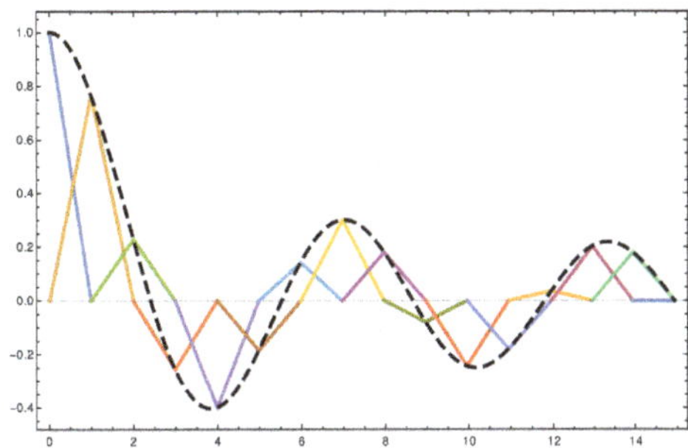

Interpolation of a Bessel function: 16 scaled and shifted triangular basis functions (colors) used to reconstruct a zeroeth order Bessel function J_o (black).

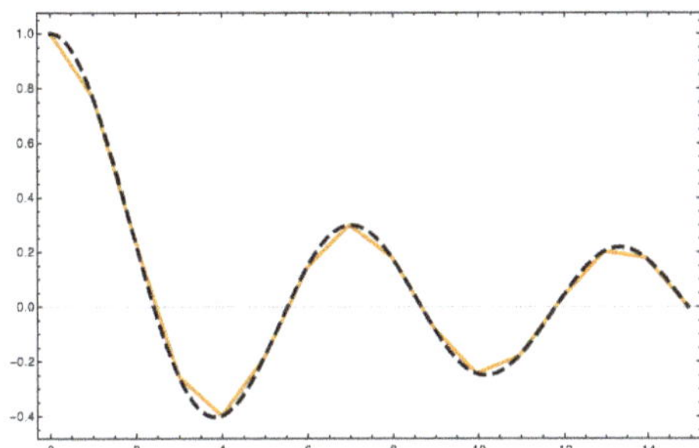

The linear combination of basis functions (yellow) reproduces J_o (blue) to any desired accuracy.

To complete the discretization, we must select a basis of V. In the one-dimensional case, for each control point x_k we will choose the piecewise linear function v_k in V whose value is 1 at x_k and zero at every $x_j, j \neq k$, i.e.,

$$v_k(x) = \begin{cases} \dfrac{x - x_{k-1}}{x_k - x_{k-1}} & \text{if } x \in [x_{k-1}, x_k], \\ \dfrac{x_{k+1} - x}{x_{k+1} - x_k} & \text{if } x \in [x_k, x_{k+1}], \\ 0 & \text{otherwise,} \end{cases}$$

for $k = 1, \ldots, n$ this basis is a shifted and scaled tent function. For the two-dimensional case, we choose again one basis function v_k per vertex x_k of the triangulation of the planar region Ω. The function v_k is the unique function of V whose value is 1 at x_k and zero at every $x_j, j \neq k$.

Depending on the author, the word "element" in "finite element method" refers either to the triangles in the domain, the piecewise linear basis function, or both. So for instance, an author interested in curved domains might replace the triangles with curved primitives, and so might describe the elements as being curvilinear. On the other hand, some authors replace "piecewise linear" by "piecewise quadratic" or even "piecewise polynomial". The author might then say "higher order element" instead of "higher degree polynomial". Finite element method is not restricted to triangles (or tetrahedra in 3-d, or higher order simplexes in multidimensional spaces), but can be defined on quadrilateral subdomains (hexahedra, prisms, or pyramids in 3-d, and so on). Higher order shapes (curvilinear elements) can be defined with polynomial and even non-polynomial shapes (e.g. ellipse or circle).

Examples of methods that use higher degree piecewise polynomial basis functions are the hp-FEM and spectral FEM.

More advanced implementations (adaptive finite element methods) utilize a method to assess the quality of the results (based on error estimation theory) and modify the mesh during the solution aiming to achieve approximate solution within some bounds from the exact solution of the continuum problem. Mesh adaptivity may utilize various techniques, the most popular are:

- Moving nodes (r-adaptivity).
- Refining (and unrefining) elements (h-adaptivity).
- Changing order of base functions (p-adaptivity).
- Combinations of the above (hp-adaptivity).

Small Support of the Basis

Solving the two-dimensional problem in the disk centered at the origin and radius 1, with zero boundary conditions.

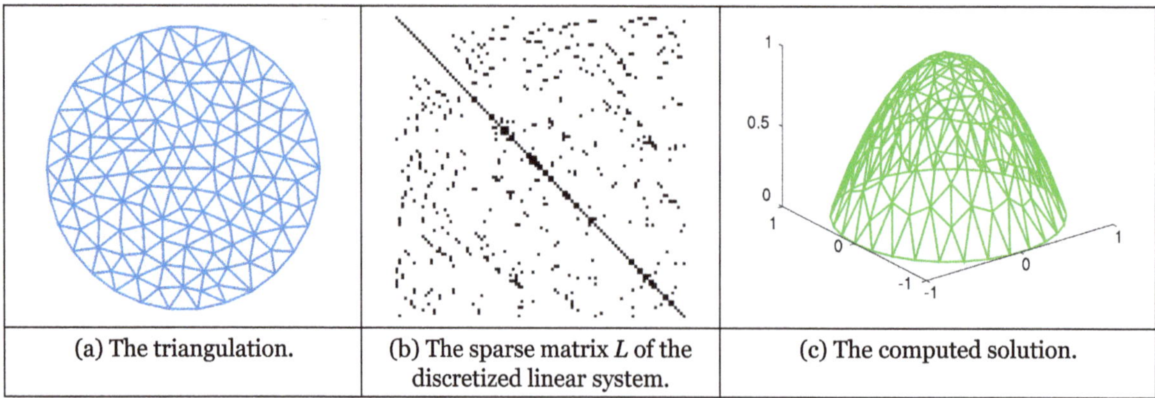

(a) The triangulation. (b) The sparse matrix L of the discretized linear system. (c) The computed solution.

The primary advantage of this choice of basis is that the inner products:

$$\langle v_j, v_k \rangle = \int_0^1 v_j v_k \, dx$$

and

$$\phi(v_j, v_k) = \int_0^1 v_{j'} v_{k'} \, dx$$

will be zero for almost all j,k. (The matrix containing $\langle v_j, v_k\rangle$ in the (j,k) location is known as the Gramian matrix.) In the one dimensional case, the support of v_k is the interval $[x_{k-1}, x_{k+1}]$. Hence, the integrands of $\langle v_j, v_k\rangle$ and $\phi(v_j, v_k)$ are identically zero whenever $|j-k|>1$.

Similarly, in the planar case, if x_j and x_k do not share an edge of the triangulation, then the integrals:

$$\int_\Omega v_j v_k \, ds$$

and

$$\int_\Omega \nabla v_j \cdot \nabla v_k \, ds$$

are both zero.

Matrix Form of the Problem

If we write $u(x) = \sum_{k=1}^{n} u_k v_k(x)$ and $f(x) \quad \sum f_k v_k(x)$ then problem taking $v(x) = v_j(x)$ for $j = 1,\ldots,n$, becomes:

$$-\sum_{k=1}^{n} u_k \phi(v_k, v_j) = \sum_{k=1}^{n} f_k \int v_k v_j dx \text{ for } j = 1,\ldots,n.$$

If we denote by u and f the column vectors $(u_1,\ldots,u_n)^t$ and $(f_1,\ldots,f_n)^t$ and if we let,

$$L = (L_{ij})$$

and

$$M = (M_{ij})$$

be matrices whose entries are,

$$L_{ij} \phi(v_i, v_j)$$

and

$$M_{ij} = \int v_i v_j dx$$

then we may rephrase as:

$$-\mathrm{Lu} = \mathrm{Mf}.$$

It is not necessary to assume $f(x) = \sum_{k=1}^{n} f_k v_k(x)$. For a general function $f = (x)$ problem with $v(x) = v_j(x)$ for $j = 1,\ldots,n$ becomes actually simpler, since no matrix M is used,

$$-\mathrm{Lu} = \mathrm{b}$$

where $b = (b_1, \ldots, b_n)^t$ and $b_j = \int f v_j dx$ for $j = 1, \ldots, n$.

As we have discussed before, most of the entries of L and M are zero because the basis functions v_k have small support. So we now have to solve a linear system in the unknown u where most of the entries of the matrix L which we need to invert, are zero.

Such matrices are known as sparse matrices, and there are efficient solvers for such problems (much more efficient than actually inverting the matrix). In addition, L is symmetric and positive definite, so a technique such as the conjugate gradient method is favored. For problems that are not too large, sparse LU decompositions and Cholesky decompositions still work well. For instance, MATLAB's backslash operator (which uses sparse LU, sparse Cholesky, and other factorization methods) can be sufficient for meshes with a hundred thousand vertices.

The matrix L is usually referred to as the stiffness matrix, while the matrix M is dubbed the mass matrix.

General Form of the Finite Element Method

In general, the finite element method is characterized by the following process:

- One chooses a grid for Ω. In the preceding treatment, the grid consisted of triangles, but one can also use squares or curvilinear polygons.
- Then, one chooses basis functions. In our discussion, we used piecewise linear basis functions, but it is also common to use piecewise polynomial basis functions.

A separate consideration is the smoothness of the basis functions. For second order elliptic boundary value problems, piecewise polynomial basis function that are merely continuous suffice (i.e., the derivatives are discontinuous.) For higher order partial differential equations, one must use smoother basis functions. For instance, for a fourth order problem such as $u_{xxxx} + u_{yyyy} = f$ one may use piecewise quadratic basis functions that are C^1.

Another consideration is the relation of the finite-dimensional space V to its infinite-dimensional counterpart, in the examples above H_0^1. A conforming element method is one in which the space V is a subspace of the element space for the continuous problem. The example above is such a method. If this condition is not satisfied, we obtain a nonconforming element method, an example of which is the space of piecewise linear functions over the mesh which are continuous at each edge midpoint. Since these functions are in general discontinuous along the edges, this finite-dimensional space is not a subspace of the original H_0^1.

Typically, one has an algorithm for taking a given mesh and subdividing it. If the main method for increasing precision is to subdivide the mesh, one has an *h*-method (*h* is customarily the diameter of the largest element in the mesh.) In this manner, if one shows that the error with a grid h is bounded above by Ch^p for some $C < \infty$ and $p > 0$, then one has an order *p* method. Under certain hypotheses (for instance, if the domain is convex), a piecewise polynomial of order d method will have an error of order $p = d + 1$.

If instead of making *h* smaller, one increases the degree of the polynomials used in the basis function, one has a *p*-method. If one combines these two refinement types, one obtains an *hp*-method (hp-FEM). In the hp-FEM, the polynomial degrees can vary from element to element. High order methods with large uniform p are called spectral finite element methods (SFEM). These are not to be confused with spectral methods.

For vector partial differential equations, the basis functions may take values in $\mathbb{R}^n$.

Various Types of Finite Element Methods

AEM

The Applied Element Method, or AEM combines features of both FEM and Discrete element method, or (DEM).

Generalized Finite Element Method

The generalized finite element method (GFEM) uses local spaces consisting of functions, not necessarily polynomials, that reflect the available information on the unknown solution and thus ensure good local approximation. Then a partition of unity is used to "bond" these spaces together to form the approximating subspace. The effectiveness of GFEM has been shown when applied to problems with domains having complicated boundaries, problems with micro-scales, and problems with boundary layers.

Mixed Finite Element Method

The mixed finite element method is a type of finite element method in which extra independent variables are introduced as nodal variables during the discretization of a partial differential equation problem.

hp-FEM

The hp-FEM combines adaptively, elements with variable size h and polynomial degree p in order to achieve exceptionally fast, exponential convergence rates.

hpk-FEM

The hpk-FEM combines adaptively, elements with variable size h, polynomial degree of the local approximations p and global differentiability of the local approximations $(k-1)$ in order to achieve best convergence rates.

XFEM

The extended finite element method (XFEM) is a numerical technique based on the generalized finite element method (GFEM) and the partition of unity method (PUM). It extends the classical finite element method by enriching the solution space for solutions to differential equations with discontinuous functions. Extended finite element methods enrich the approximation space so that it is able to naturally reproduce the challenging feature associated with the problem of interest: the discontinuity, singularity, boundary layer, etc. It was shown that for some problems, such an embedding of the problem's feature into the approximation space can significantly improve convergence rates and accuracy. Moreover, treating problems with discontinuities with XFEMs suppresses the need to mesh and remesh the discontinuity surfaces, thus alleviating the computational costs and projection errors associated with conventional finite element methods, at the cost of restricting the discontinuities to mesh edges.

Several research codes implement this technique to various degrees: 1. GetFEM++ 2. xfem++ 3. openxfem++ XFEM has also been implemented in codes like Altair Radioss, ASTER, Morfeo and Abaqus. It is increasingly being adopted by other commercial finite element software, with a few plugins and actual core implementations available (ANSYS, SAMCEF, OOFELIE, etc.).

Scaled Boundary Finite Element Method (SBFEM)

The introduction of the scaled boundary finite element method (SBFEM) came from Song and Wolf. The SBFEM has been one of the most profitable contributions in the area of numerical analysis of fracture mechanics problems. It is a semi-analytical fundamental-solutionless method which combines the advantages of both the finite element formulations and procedures, and the boundary element discretization. However, unlike the boundary element method, no fundamental differential solution is required.

S-FEM

The S-FEM, Smoothed Finite Element Methods are a particular class of numerical simulation algorithms for the simulation of physical phenomena. It was developed by combining meshfree methods with the finite element method.

Spectral Element Method

Spectral element methods combine the geometric flexibility of finite elements and the acute accuracy of spectral methods. Spectral methods are the approximate solution of weak form partial equations that are based on high-order Lagragian interpolants and used only with certain quadrature rules.

Link with the Gradient Discretisation Method

Some types of finite element methods (conforming, nonconforming, mixed finite element methods) are particular cases of the gradient discretisation method (GDM). Hence the convergence properties of the GDM, which are established for a series of problems (linear and non linear elliptic problems, linear, nonlinear and degenerate parabolic problems), hold as well for these particular finite element methods.

Comparison to the Finite Difference Method

The finite difference method (FDM) is an alternative way of approximating solutions of PDEs. The differences between FEM and FDM are:

- The most attractive feature of the FEM is its ability to handle complicated geometries (and boundaries) with relative ease. While FDM in its basic form is restricted to handle rectangular shapes and simple alterations thereof, the handling of geometries in FEM is theoretically straightforward.

- FDM is not usually used for irregular CAD geometries but more often rectangular or block shaped models.
- The most attractive feature of finite differences is that it is very easy to implement.
- There are several ways one could consider the FDM a special case of the FEM approach. E.g., first order FEM is identical to FDM for Poisson's equation, if the problem is discretized by a regular rectangular mesh with each rectangle divided into two triangles.
- There are reasons to consider the mathematical foundation of the finite element approximation more sound, for instance, because the quality of the approximation between grid points is poor in FDM.
- The quality of a FEM approximation is often higher than in the corresponding FDM approach, but this is extremely problem-dependent and several examples to the contrary can be provided.

Generally, FEM is the method of choice in all types of analysis in structural mechanics (i.e. solving for deformation and stresses in solid bodies or dynamics of structures) while computational fluid dynamics (CFD) tends to use FDM or other methods like finite volume method (FVM). CFD problems usually require discretization of the problem into a large number of cells/gridpoints (millions and more), therefore cost of the solution favors simpler, lower order approximation within each cell. This is especially true for 'external flow' problems, like air flow around the car or airplane, or weather simulation.

Application

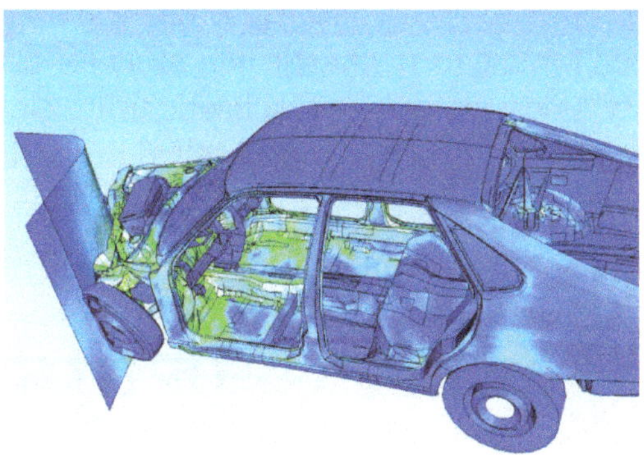

Visualization of how a car deforms in an asymmetrical crash using finite element analysis.

A variety of specializations under the umbrella of the mechanical engineering discipline (such as aeronautical, biomechanical, and automotive industries) commonly use integrated FEM in design and development of their products. Several modern FEM packages include specific components such as thermal, electromagnetic, fluid, and structural working environments. In a structural simulation, FEM helps tremendously in producing stiffness and strength visualizations and also in minimizing weight, materials, and costs.

FEM allows detailed visualization of where structures bend or twist, and indicates the distribution of stresses and displacements. FEM software provides a wide range of simulation options for controlling the complexity of both modeling and analysis of a system. Similarly, the desired level of accuracy required and associated computational time requirements can be managed simultaneously to address most engineering applications. FEM allows entire designs to be constructed, refined, and optimized before the design is manufactured. The mesh is an integral part of the model and it must be controlled carefully to give the best results. Generally the higher the number of elements in a mesh, the more accurate the solution of the discretized problem. However, there is a value at which the results converge and further mesh refinement does not increase accuracy.

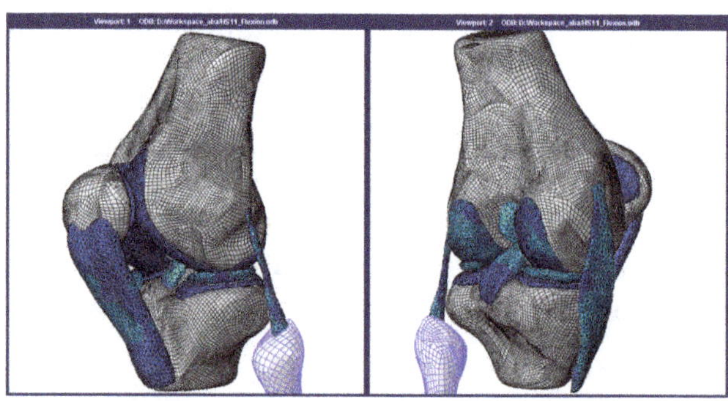

Finite Element Model of a human knee joint.

This powerful design tool has significantly improved both the standard of engineering designs and the methodology of the design process in many industrial applications. The introduction of FEM has substantially decreased the time to take products from concept to the production line. It is primarily through improved initial prototype designs using FEM that testing and development have been accelerated. In summary, benefits of FEM include increased accuracy, enhanced design and better insight into critical design parameters, virtual prototyping, fewer hardware prototypes, a faster and less expensive design cycle, increased productivity, and increased revenue.

In the 1990s FEA was proposed for use in stochastic modeling for numerically solving probability models and later for reliability assessment. The stochastic finite element method has since been applied to many branches of engineering, often being applied to characterize variability in material properties.

MULTIDISCIPLINARY DESIGN OPTIMIZATION

Multidisciplinary design optimization (MDO) is a field of engineering that uses optimization methods to solve design problems incorporating a number of disciplines. It is also known as multidisciplinary system design optimization (MSDO).

MDO allows designers to incorporate all relevant disciplines simultaneously. The optimum of the simultaneous problem is superior to the design found by optimizing each discipline sequentially, since it can exploit the interactions between the disciplines. However, including all disciplines simultaneously significantly increases the complexity of the problem.

These techniques have been used in a number of fields, including automobile design, naval architecture, electronics, architecture, computers, and electricity distribution. However, the largest number of applications has been in the field of aerospace engineering, such as aircraft and spacecraft design. For example, the proposed Boeing blended wing body (BWB) aircraft concept has used MDO extensively in the conceptual and preliminary design stages. The disciplines considered in the BWB design are aerodynamics, structural analysis, propulsion, control theory, and economics.

Traditionally engineering has normally been performed by teams, each with expertise in a specific discipline, such as aerodynamics or structures. Each team would use its members' experience and judgement to develop a workable design, usually sequentially. For example, the aerodynamics experts would outline the shape of the body, and the structural experts would be expected to fit their design within the shape specified. The goals of the teams were generally performance-related, such as maximum speed, minimum drag, or minimum structural weight.

Between 1970 and 1990, two major developments in the aircraft industry changed the approach of aircraft design engineers to their design problems. The first was computer-aided design, which allowed designers to quickly modify and analyse their designs. The second was changes in the procurement policy of most airlines and military organizations, particularly the military of the United States, from a performance-centred approach to one that emphasized lifecycle cost issues. This led to an increased concentration on economic factors and the attributes known as the "ilities" including manufacturability, reliability, maintainability, etc.

Since 1990, the techniques have expanded to other industries. Globalization has resulted in more distributed, decentralized design teams. The high-performance personal computer has largely replaced the centralized supercomputer and the Internet and local area networks have facilitated sharing of design information. Disciplinary design software in many disciplines (such as OptiStruct or NASTRAN, a finite element analysis program for structural design) have become very mature. In addition, many optimization algorithms, in particular the population-based algorithms, have advanced significantly.

Origins in Structural Optimization

Whereas optimization methods are nearly as old as calculus, dating back to Isaac Newton, Leonhard Euler, Daniel Bernoulli, and Joseph Louis Lagrange, who used them to solve problems such as the shape of the catenary curve, numerical optimization reached prominence in the digital age. Its systematic application to structural design dates to its advocacy by Schmit in 1960. The success of structural optimization in the 1970s motivated the emergence of multidisciplinary design optimization (MDO) in the 1980s. Jaroslaw Sobieski championed decomposition methods specifically designed for MDO applications. The following synopsis focuses on optimization methods for MDO.

Gradient-Based Methods

There were two schools of structural optimization practitioners using gradient-based methods during the 1960s and 1970s: optimality criteria and mathematical programming. The optimality criteria school derived recursive formulas based on the Karush–Kuhn–Tucker (KKT) necessary conditions for an optimal design. The KKT conditions were applied to classes of structural problems such as minimum weight design with constraints on stresses, displacements, buckling, or frequencies to derive resizing expressions particular to each class. The mathematical programming school employed classical gradient-based methods to structural optimization problems. The method of usable feasible directions, Rosen's gradient projection (generalized reduce gradient) method, sequential unconstrained minimization techniques, sequential linear programming and eventually sequential quadratic programming methods were common choices. Schittkowski et al. reviewed the methods current by the early 1990s.

The gradient methods unique to the MDO community derive from the combination of optimality criteria with math programming, first recognized in the seminal work of Fleury and Schmit who constructed a framework of approximation concepts for structural optimization. They recognized that optimality criteria were so successful for stress and displacement constraints, because that approach amounted to solving the dual problem for Lagrange multipliers using linear Taylor series approximations in the reciprocal design space. In combination with other techniques to improve efficiency, such as constraint deletion, regionalization, and design variable linking, they succeeded in uniting the work of both schools. This approximation concepts based approach forms the basis of the optimization modules in modern structural design software such as Altair – Optistruct, ASTROS, MSC.Nastran, PHX ModelCenter, Genesis, iSight, and I-DEAS.

Approximations for structural optimization were initiated by the reciprocal approximation Schmit and Miura for stress and displacement response functions. Other intermediate variables were employed for plates. Combining linear and reciprocal variables, Starnes and Haftka developed a conservative approximation to improve buckling approximations. Fadel chose an appropriate intermediate design variable for each function based on a gradient matching condition for the previous point. Vanderplaats initiated a second generation of high quality approximations when he developed the force approximation as an intermediate response approximation to improve the approximation of stress constraints. Canfield developed a Rayleigh quotient approximation to improve the accuracy of eigenvalue approximations.

Non-Gradient-Based Methods

In recent years, non-gradient-based evolutionary methods including genetic algorithms, simulated annealing, and ant colony algorithms came into existence. At present, many researchers are striving to arrive at a consensus regarding the best modes and methods for complex problems like impact damage, dynamic failure, and real-time analyses. For this purpose, researchers often employ multiobjective and multicriteria design methods.

Recent MDO Methods

MDO practitioners have investigated optimization methods in several broad areas in the last dozen years. These include decomposition methods, approximation methods, evolutionary algorithms, memetic algorithms, response surface methodology, reliability-based optimization, and multi-objective optimization approaches.

The exploration of decomposition methods has continued in the last dozen years with the development and comparison of a number of approaches, classified variously as hierarchic and non hierarchic, or collaborative and non collaborative. Approximation methods spanned a diverse set of approaches, including the development of approximations based on surrogate models (often referred to as metamodels), variable fidelity models, and trust region management strategies. The development of multipoint approximations blurred the distinction with response surface methods. Some of the most popular methods include Kriging and the moving least squares method.

Response surface methodology, developed extensively by the statistical community, received much attention in the MDO community in the last dozen years. A driving force for their use has been the development of massively parallel systems for high performance computing, which are naturally suited to distributing the function evaluations from multiple disciplines that are required for the construction of response surfaces. Distributed processing is particularly suited to the design process of complex systems in which analysis of different disciplines may be accomplished naturally on different computing platforms and even by different teams.

Evolutionary methods led the way in the exploration of non-gradient methods for MDO applications. They also have benefited from the availability of massively parallel high performance computers, since they inherently require many more function evaluations than gradient-based methods. Their primary benefit lies in their ability to handle discrete design variables and the potential to find globally optimal solutions.

Reliability-based optimization (RBO) is a growing area of interest in MDO. Like response surface methods and evolutionary algorithms, RBO benefits from parallel computation, because the numeric integration to calculate the probability of failure requires many function evaluations. One of the first approaches employed approximation concepts to integrate the probability of failure. The classical first-order reliability method (FORM) and second-order reliability method (SORM) are still popular. Professor Ramana Grandhi used appropriate normalized variables about the most probable point of failure, found by a two-point adaptive nonlinear approximation to improve the accuracy and efficiency. Southwest Research Institute has figured prominently in the development of RBO, implementing state-of-the-art reliability methods in commercial software. RBO has reached sufficient maturity to appear in commercial structural analysis programs like Altair's Optistruct and MSC's Nastran.

Utility-based probability maximization was developed in response to some logical concerns (e.g., Blau's Dilemma) with reliability-based design optimization. This approach focuses on maximizing the joint probability of both the objective function exceeding some value and of all the constraints being satisfied. When there is no objective function, utility-based probability maximization reduces to a probability-maximization problem. When there are no uncertainties in the constraints, it reduces to a constrained utility-maximization problem. (This second equivalence arises because the utility of a function can always be written as the probability of that function exceeding some random variable.) Because it changes the constrained optimization problem associated with reliability-based optimization into an unconstrained optimization problem, it often leads to computationally more tractable problem formulations.

In the marketing field there is a huge literature about optimal design for multiattribute products and services, based on experimental analysis to estimate models of consumers' utility functions. These methods are known as Conjoint Analysis. Respondents are presented with alternative products, measuring preferences about the alternatives using a variety of scales and the utility function is estimated with different methods (varying from regression and surface response methods to choice models). The best design is formulated after estimating the model. The experimental design is usually optimized to minimize the variance of the estimators. These methods are widely used in practice.

Problem Formulation

Problem formulation is normally the most difficult part of the process. It is the selection of design variables, constraints, objectives, and models of the disciplines. A further consideration is the strength and breadth of the interdisciplinary coupling in the problem.

Design Variables

A design variable is a specification that is controllable from the point of view of the designer. For instance, the thickness of a structural member can be considered a design variable. Another might be the choice of material. Design variables can be continuous (such as a wing span), discrete (such as the number of ribs in a wing), or boolean (such as whether to build a monoplane or a biplane). Design problems with continuous variables are normally solved more easily. Design variables are often bounded, that is, they often have maximum and minimum values. Depending on the solution method, these bounds can be treated as constraints or separately.

One of the important variables that needs to be accounted is an uncertainty. Uncertainty, often referred to as epistemic uncertainty, arises due to lack of knowledge or incomplete information. Uncertainty is essentially unknown variable but it may causes the failure of system.

Constraints

A constraint is a condition that must be satisfied in order for the design to be feasible. An example of a constraint in aircraft design is that the lift generated by a wing must be equal to the weight of the aircraft. In addition to physical laws, constraints can reflect resource limitations, user requirements, or bounds on the validity of the analysis models. Constraints can be used explicitly by the solution algorithm or can be incorporated into the objective using Lagrange multipliers.

Objectives

An objective is a numerical value that is to be maximized or minimized. For example, a designer may wish to maximize profit or minimize weight. Many solution methods work only with single objectives. When using these methods, the designer normally weights the various objectives and sums them to form a single objective. Other methods allow multi objective optimization, such as the calculation of a Pareto front.

Models

The designer must also choose models to relate the constraints and the objectives to the design variables. These models are dependent on the discipline involved. They may be empirical models, such as a regression analysis of aircraft prices, theoretical models, such as from computational fluid dynamics, or reduced-order models of either of these. In choosing the models the designer must trade off fidelity with analysis time.

The multidisciplinary nature of most design problems complicates model choice and implementation. Often several iterations are necessary between the disciplines in order to find the values of the objectives and constraints. As an example, the aerodynamic loads on a wing affect the structural deformation of the wing. The structural deformation in turn changes the shape of the wing and the aerodynamic loads. Therefore, in analysing a wing, the aerodynamic and structural analyses must be run a number of times in turn until the loads and deformation converge.

Standard Form

Once the design variables, constraints, objectives, and the relationships between them have been chosen, the problem can be expressed in the following form:

$$\text{find } \mathrm{x} \text{ that minimizes } J(\mathrm{x}) \text{ subject to g } \mathrm{g}(\mathrm{x}) \leq 0,\ \mathrm{h}(\mathrm{x}) = 0 \text{ and } \mathrm{x}_{lb} \leq \mathrm{x} \leq \mathrm{x}_{ub}$$

where J is an objective, x is a vector of design variables, g is a vector of inequality constraints, h is a vector of equality constraints, and x_{lb} and x_{ub} are vectors of lower and upper bounds on the design variables. Maximization problems can be converted to minimization problems by multiplying the objective by -1. Constraints can be reversed in a similar manner. Equality constraints can be replaced by two inequality constraints.

Problem Solution

The problem is normally solved using appropriate techniques from the field of optimization. These include gradient-based algorithms, population-based algorithms, or others. Very simple problems can sometimes be expressed linearly; in that case the techniques of linear programming are applicable.

Gradient-Based Methods

- Adjoint equation
- Newton's method
- Steepest descent
- Conjugate gradient
- Sequential quadratic programming

Gradient-Free Methods

- Hooke-Jeeves pattern search
- Nelder-Mead method

Population-Based Methods

- Genetic algorithm
- Memetic algorithm
- Particle swarm optimization

- Harmony search
- ODMA

Other Methods

- Random search
- Grid search
- Simulated annealing
- Direct search
- IOSO (Indirect Optimization based on Self-Organization)

Most of these techniques require large numbers of evaluations of the objectives and the constraints. The disciplinary models are often very complex and can take significant amounts of time for a single evaluation. The solution can therefore be extremely time-consuming. Many of the optimization techniques are adaptable to parallel computing. Much current research is focused on methods of decreasing the required time.

Also, no existing solution method is guaranteed to find the global optimum of a general problem. Gradient-based methods find local optima with high reliability but are normally unable to escape a local optimum. Stochastic methods, like simulated annealing and genetic algorithms, will find a good solution with high probability, but very little can be said about the mathematical properties of the solution. It is not guaranteed to even be a local optimum. These methods often find a different design each time they are run.

COMPUTATIONAL FLUID DYNAMICS

When an engineer is tasked with designing a new product, e.g. a winning race car for the next season, aerodynamics plays an important role in the engineering process. However, aerodynamic processes are not easily quantifiable during the concept phase. Usually the only way for the engineer to optimize his designs is to conduct physical tests on product prototypes. With the rise of computers and ever-growing computational power, the field of Computational Fluid Dynamics became a commonly applied tool for generating solutions for fluid flows with or without solid interaction. In a CFD analysis, the examination of fluid flow in accordance with its physical properties such as velocity, pressure, temperature, density and viscosity is conducted. To virtually generate a solution for a physical phenomenon associated with fluid flow, without compromise on accuracy those properties have to be considered simultaneously.

A mathematical model of the physical case and a numerical method are used in a software tool to analyze the fluid flow. For instance, the Navier-Stokes equations are specified as the mathematical model of the physical case. This describes changes on all those physical properties for both fluid flow and heat transfer. The mathematical model varies in accordance with the content of the problem such as heat transfer, mass transfer, phase change, chemical reaction, etc. Moreover, the reliability of a CFD analysis highly depends on the whole structure of the process. The verification of the mathematical model is extremely important to create an accurate case for solving the problem. Besides, the determination of proper numerical methods to generate a path through the solution is as important as a mathematical model. The software, which the analysis is conducted with is one of the key elements in generating a sustainable product development process, as the amount of physical prototypes can be reduced drastically.

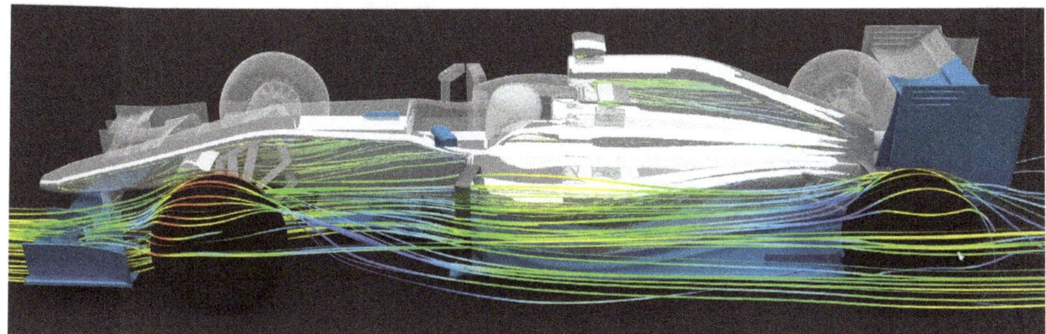

Steamlines around an F1 car.

Governing Equations

The main structure of thermo-fluids examinations is directed by governing equations that are based on the conservation law of fluid's physical properties. The basic equations are the three physics laws of conservation:

- Conservation of Mass: Continuity Equation.
- Conservation of Momentum: Momentum Equation of Newton's Second Law.
- Conservation of Energy: First Law of Thermodynamics or Energy Equation.

These principles state that mass, momentum and energy are stable constants within a closed system. Basically, what comes in, must also go out somewhere else.

The investigation of fluid flow with thermal changes relies on certain physical properties. The three unknowns which must be obtained simultaneously from these three basic conservation equations are the velocity $\vec{v}$, pressure p and the absolute temperature T. Yet p and T are considered the two required independent thermodynamics variables. The final form of the conservation equations also contains four other thermodynamics variables; density ρ the enthalpy h as well as viscosity μ and thermal conductivity k; the last two are also transport properties. Since p and T are considered two required independent thermodynamics variables, these four properties are uniquely determined by the value of p and T. Fluid flow should be analyzed to know vecv, p and T throughout every point of ... w regime. This is most important before designing any product which involves fluid flo... re, the method of fluid flow observation based on kinematic properties is a funda-
...ement of fluid can be investigated with either Lagrangian or Eulerian methods.
...ion of fluid motion is based on the theory to follow a fluid particle which is large
...perties. Initial coordinates at time t_0 and coordinates of the same particle at
...ined. To follow millions of separate particles through the path is almost im-
... method, any specific particle across the path is not followed, instead, the

velocity field as a function of time and position is examined. This missile example precisely fits to emphasize the methods.

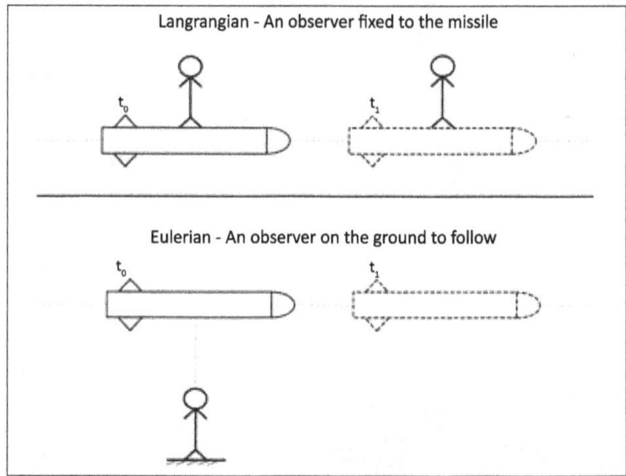

Observation of fluid motion with the methods of Lagrange and Euler.

1. Langarian: We take up every point at the beginning of the domain and trace its path till it reaches the end.

2. Eulerian: We consider a window (Control Volume) within the fluid and analyse the particle flow within this Volume.

Lagrangian formulation of motion is always time-dependent. As a, b and c are the initial coordinates of a particle; x, y, and z are coordinates of the same particle at time t. Description of motion for Lagrangian:

$$x = x(a,b,c,t) y = y(a,b,c,t) z = z(a,b,c,t)$$

In the Eulerian method, uu, vv and ww are the components of velocity at the point (x, y, z) (x, y, z) at the time t. Thus, u, v and w are the unknowns which are functions of the independent variables x, y, z and t. Description of motion for Eulerian for any particular value of t:

$$u = u(x, y, z, t)$$
$$v = v(x, y, z, t)$$
$$w = w(x, y, z, t)$$

Conservation of Mass is specified as equation below:

$$\frac{D\rho}{Dt} + \rho(\nabla \cdot \vec{v}) = 0$$

where ρ is the density, $\vec{v}$ the velocity and ∇ the gradient operator.

$$\vec{\nabla} = \vec{i}\frac{\partial}{\partial x} + \vec{j}\frac{\partial}{\partial y} + \vec{k}\frac{\partial}{\partial z}$$

If the density is constant, the flow is assumed to be incompressible and then continuity reduces it to:

$$\frac{D\rho}{Dt}=0\rightarrow\nabla\cdot\vec{v}=\frac{\partial u}{\partial x}+\frac{\partial v}{\partial y}+\frac{\partial w}{\partial z}=0$$

Conservation of Momentum which can be specified as Navier-Stokes Equation:

$$\overbrace{\frac{\partial}{\partial t}(\rho\vec{v})}^{I}+\overbrace{\nabla.(\rho\vec{v}\vec{v})}^{II}=\overbrace{-\nabla p}^{III}+\overbrace{\nabla.\left(\bar{\bar{\tau}}\right)}^{IV}+\overbrace{\rho\vec{g}}^{V}$$

where static pressure p, viscous stress tensor $\bar{\bar{\tau}}$ and gravitational force $\rho\vec{g}$.

- Local change with time
- Momentum convection
- Surface force
- Diffusion term
- Mass force

Viscous stress tensor $\bar{\bar{\tau}}$ can be specified as below in accordance with Stoke's Hypothesis:

$$\tau ij=\mu\frac{\partial vi}{\partial xj}+\frac{\partial vj}{\partial xi}-\frac{2}{3}(\nabla\cdot\vec{v}\,)\delta_{ij}$$

If the fluid is assumed to be of constant density (can't be compressed), the equations are greatly simplified that viscosity coefficient μ is assumed constant. Therefore, many terms vanished through equation results in a much simpler Navier-Stokes Equation:

$$\rho\frac{D\vec{v}}{Dt}=-\nabla p+\mu\nabla^2\vec{v}+\rho\vec{g}$$

Conservation of Energy is the first law of thermodynamics which states that the sum of the work and heat added to the system will result in the increase of the energy in the system:

$$dE_t=dQ+dW$$

where dQ is the heat added to the system, dW is the work done on the system and dE_t is the increment in the total energy of the system. One of the common types of an energy equation is:

$$\rho\left[\overbrace{\frac{\partial h}{\partial t}}^{I}+\overbrace{\nabla.(h\vec{v})}^{II}\right]=\overbrace{-\frac{\partial p}{\partial t}}^{III}+\overbrace{\nabla\cdot(k\nabla T)}^{IV}+\overbrace{\phi}^{V}$$

Partial Differential Equations (PDEs)

The Mathematical model merely gives us interrelation between the transport parameters which are involved in the whole process, either directly or indirectly. Even though every single term in those equations has relative effect on the physical phenomenon, changes in parameters should be considered simultaneously through the numerical solution which comprises differential equations, vector and tensor notations. A PDE comprises more than one variable and their derivation which is specified with "∂" instead of "d". If the derivation of the equation is conducted with "d", these equations are called as Ordinary Differential Equations (ODE) that contains a single variable and its derivation. The PDEs are implicated to transform the differential operator (∂) into an algebraic operator in order to get a solution. Heat transfer, fluid dynamics, acoustic, electronics and quantum mechanics are the fields that PDEs are highly used to generate solutions.

Example of ODE:

$$\frac{d^2x}{dt^2} = x \to x(t) \text{ where T is the single variable.}$$

Example of PDE:

$$\frac{\partial f}{\partial x} + \frac{\partial f}{\partial y} = 5 \to f(x, y) \text{ where both x and y are the variables.}$$

What is the significance of PDEs to seeking a solution on governing equations? To answer this question, we initially examine the basic structure of some PDEs as to create connotation.

For instance:

$$\frac{\partial^2 f}{\partial x^2} + \frac{\partial^2 f}{\partial y^2} = 0 \to f(x, y) \to \text{Laplace Equation}$$

Comparison between equation ($\frac{D\rho}{Dt} = 0 \to \nabla \cdot \vec{v} = \frac{\partial u}{\partial x} + \frac{\partial v}{\partial y} + \frac{\partial w}{\partial z} = 0$) and equation $\frac{\partial^2 f}{\partial x^2} + \frac{\partial^2 f}{\partial y^2} = 0 \to f(x, y) \to \text{Laplace Equation}$ specifies the Laplace part of the continuity equation. What is the next step? What does this Laplace analogy mean? To start solving these enormous equations, the next step comes through discretization to ignite the numerical solution process. The numerical solution is a discretization-based method used in order to obtain approximate solutions to complex problems which cannot be solved with analytic methods because of complexity and ambiguities. Solution processes without discretization merely give you an analytic solution which is exact but simple. Moreover, the accuracy of the numerical solution highly depends on the quality of the discretization. Broadly used discretization methods might be specified such as finite difference, finite volume, finite element, spectral (element) methods and boundary element.

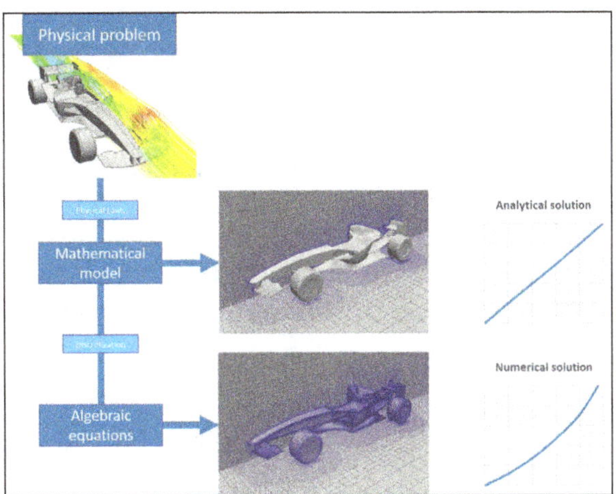

The panoramic structure of a CFD project and its stages.

Mesh Convergence

Multitasking is one of the plagues of the century that generally ends up with procrastination or failure. Therefore, having planned, segmented and sequenced tasks is much more appropriate to achieving goals: this has also been working for CFD. In order to conduct an analysis, the solution domain is split into multiple sub-domains which are called cells. The combination of these cells in the computational structure is named mesh.

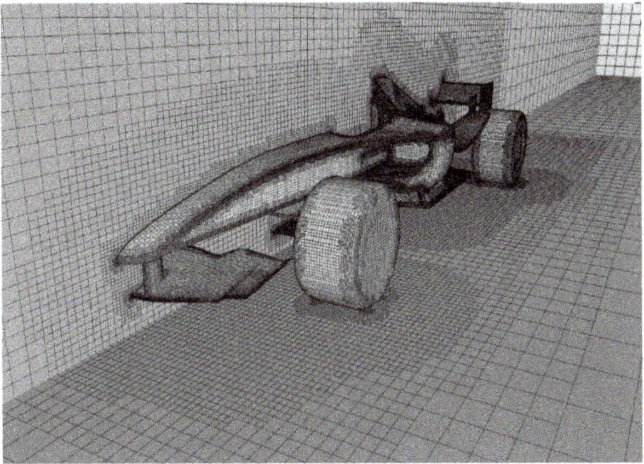

Mesh of a Formula 1 car.

The mesh as simplification of the domain is needed, because it is only possible to solve the mathematical model under the assumption of linearity. This means that we need to ensure that the behavior of the variables we want to solve for can assumed to be linear within each cell. This requirement also implies that a finer mesh (generated via mesh refinement steps) is needed for areas in the domain where the physical properties to be predicted are suspected to be highly volatile.

Errors based on mesh structure are an often encountered issue which results in the failure of the simulation. This might happen because the mesh is too coarse and doesn't cover all effects that happen in this single element one by one, but rather cover multiple effects that then change as the mesh gets finer. Therefore, a study of independency needs to be carried out. The accuracy of the solution enormously depends on the mesh structure. To conduct accurate solutions and obtaining reliable results, the analyst has to be extremely careful on the type of cell, the number of cell and the computation time.

The optimization of those restrictions is defined as mesh convergence which might be sorted as below:

1. Generate a mesh structure that has a quite low number of elements and carries out analysis. Before, assure that the mesh quality and coverage of CAD model is reasonable to examination.
2. Regenerate mesh structures with a higher number of elements. Conduct analysis again. Compare results in accordance with properties of examined case. For instance, if a case is an examination of internal flow through a channel, pressure drop at critical regions might be used as comparison.
3. Keep on ascending to a number of elements where results converge satisfactorily with previous one(s).

Therefore, errors, based on mesh structure, can be eliminated and optimum value for number of elements might virtually be achieved as to optimize calculation time and necessary computation resources. An illustration is shown in figure below that looks into static pressure change at imaginary region X through increase in number of elements. According to figure below, around 1,000,000 elements would have been sufficient to conduct a reliable study.

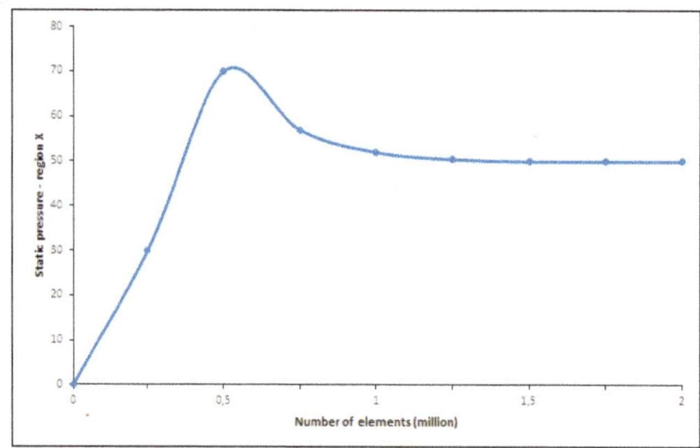

Example of mesh convergence analysis.

Convergence in Computational Fluid Dynamics

Creating a sculpture requires a highly talented artist with the ability to imagine the final product from the beginning. Yet a sculpture can be, for example, a simple piece of rock in the beginning, but might become an exceptional artwork in the end. A completely gradual processing throughout carving is an important issue to obtain the desired unique shape. Keep in mind that in every single process, some of the elements, such as stone particles, leftovers, are thrown away from the object. CFD also has a similar structure that relies on gradual processing during the analysis. In regions that are highly critical to the simulation results (for example a spoiler on a Formula 1 car) the mesh is refined into smaller elements to make the simulation more accurate.

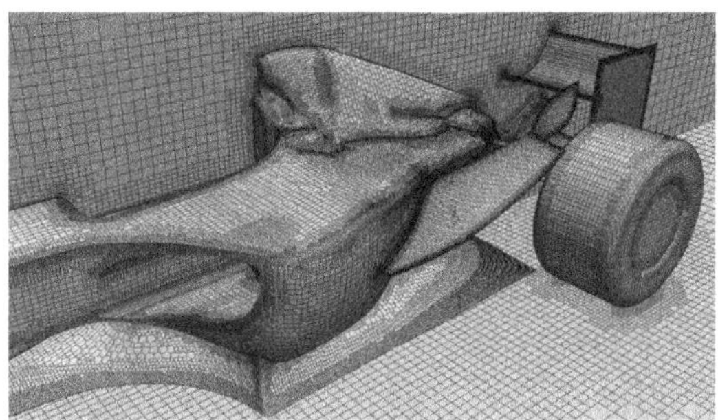

Mesh refinement example by Sim Scale.

Convergence is a major issue for computational analysis. The movement of fluid has a non-linear mathematical model with various complex models such as turbulence, phase change and mass transfer. Apart from the analytical solution, the numerical solution goes through an iterative scheme where results are obtained by the reduction of errors among previous stages. The differences between the last two values specify the error. When the absolute error is descending, the reliability of the result increases, which means that the result converges towards a stable solution.

The criteria for convergence vary with the mathematical models such as turbulence, multi-phase, etc. How do analysts decide when the solution is converged? Convergence should go on and on until a steady-state condition has been obtained, even if the aimed case is transient, which indicates results change through time. Convergence has to be realized for each time-step as if they all are a steady-state process. What are the criteria for convergence? The rate of accuracy (acceptable error), complexity of the case and calculation time have to be considered as major topics to carry out an optimal process. The residuals of equations, like stone leftovers, change over each iteration. As iterations get down to the threshold value, convergence is achieved. For a transient case, those processes have to be achieved for each of the time steps.

The convergence might be diversified as follows below:

- Can be accelerated by parameters as initial conditions, under-relaxation and Courant number.
- Doesn't always have to be correct, yet solution can converge, preferred mathematical model and mesh would be incorrect or have ambiguities.
- Can be stabilized within several methods like reasonable mesh quality, mesh refinement, using discretization schemes first- to second-order.
- Ensure that solution should be repeatable if necessary as to refrain ambiguity.

Applications of Computational Fluid Dynamics

Where there is fluid, there is CFD. Having mentioned before, the initial stage to conduct a CFD simulation is specifying an appropriate mathematical model of reality. Rapprochements and assumptions give direction through solution processes to examine the case in the computational domain. For instance, fluid flow over a sphere/cylinder is a repetitive issue that has been taught by the lecturer as an example in fluid courses. The same phenomenon is virtually available in the movement of clouds in the atmosphere which is indeed tremendou.

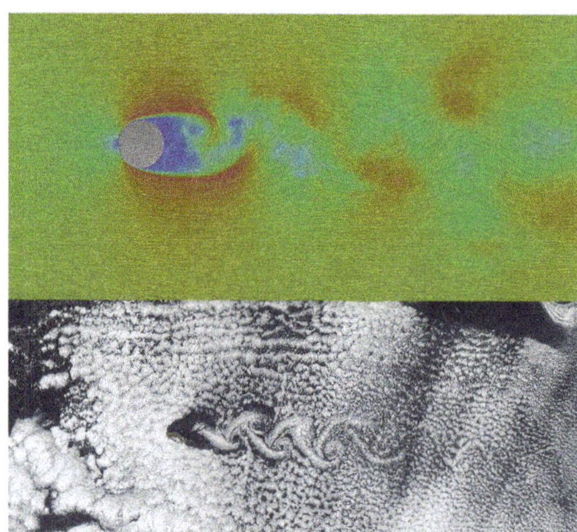

Examples of Karman vortex streets; numerical result (top) and real life example of clouds^12 (bottom).

Incompressible and Compressible Flow

If compressibility becomes a non-negligible factor, this type of analysis helps you to find solutions in a very robust and accurate way. One example would be a Large Eddy Simulation of flow around a cylinder.

Laminar and Turbulent Flow

Different turbulence models play a role in this type of analysis. A lot of computing power is required to solve turbulence simulations and its complex numerical models. The difficulty of turbulence is the simulation of changes over time. The entire domain where the simulation takes place needs to be recalculated after every time step. The analysis of a ball valve is one possible application of a turbulent flow analysis.

Mass and Thermal Transport

Mass transport simulations include smoke propagation, passive scalar transport or gas distributions. To solve these kinds of simulations, OpenFOAM solvers are used. Heat exchanger simulations are one possible application.

Different Types of CFD Applications

Computational Fluid Dynamics tools diversify in accordance with mathematical models, numerical methods, computational equipment and post-processing facilities. As a physical phenomenon could be modeled with completely different mathematical approaches, it would also be integrated with unlike numerical methods simultaneously. Thus, a conscious rapprochement is the essential factor on the path to developing CFD tools.

References

- Raj, Pradeep; Brennan, James E. (1989). "Improvements to an Euler aerodynamic method for transonic flow analysis". Journal of Aircraft. 26: 13–20. doi:10.2514/3.45717

- Computer-aided-manufacturing-beginners, products, fusion-360: autodesk.com, Retrieved 10 June, 2019
- "Spectral Element Methods". State Key Laboratory of Scientific and Engineering Computing. Retrieved 2017-07-28
- Whatiscae, general, content: simscale.com, Retrieved 25 July, 2019
- Reddy, J. N. (2006). An Introduction to the Finite Element Method (Third ed.). McGraw-Hill. ISBN 9780071267618

Permissions

All chapters in this book are published with permission under the Creative Commons Attribution Share Alike License or equivalent. Every chapter published in this book has been scrutinized by our experts. Their significance has been extensively debated. The topics covered herein carry significant information for a comprehensive understanding. They may even be implemented as practical applications or may be referred to as a beginning point for further studies.

We would like to thank the editorial team for lending their expertise to make the book truly unique. They have played a crucial role in the development of this book. Without their invaluable contributions this book wouldn't have been possible. They have made vital efforts to compile up to date information on the varied aspects of this subject to make this book a valuable addition to the collection of many professionals and students.

This book was conceptualized with the vision of imparting up-to-date and integrated information in this field. To ensure the same, a matchless editorial board was set up. Every individual on the board went through rigorous rounds of assessment to prove their worth. After which they invested a large part of their time researching and compiling the most relevant data for our readers.

The editorial board has been involved in producing this book since its inception. They have spent rigorous hours researching and exploring the diverse topics which have resulted in the successful publishing of this book. They have passed on their knowledge of decades through this book. To expedite this challenging task, the publisher supported the team at every step. A small team of assistant editors was also appointed to further simplify the editing procedure and attain best results for the readers.

Apart from the editorial board, the designing team has also invested a significant amount of their time in understanding the subject and creating the most relevant covers. They scrutinized every image to scout for the most suitable representation of the subject and create an appropriate cover for the book.

The publishing team has been an ardent support to the editorial, designing and production team. Their endless efforts to recruit the best for this project, has resulted in the accomplishment of this book. They are a veteran in the field of academics and their pool of knowledge is as vast as their experience in printing. Their expertise and guidance has proved useful at every step. Their uncompromising quality standards have made this book an exceptional effort. Their encouragement from time to time has been an inspiration for everyone.

The publisher and the editorial board hope that this book will prove to be a valuable piece of knowledge for students, practitioners and scholars across the globe.

Index